Berlin: City of Animals
The story of Germany's oldest and most famous zoo

Clemens Maier-Wolthausen

BERLIN
CITY OF ANIMALS

The story of Germany's oldest and most famous zoo

Edited by
Andreas Knieriem

English translation by
Heather Kimber and Grace Winter

Ch.Links VERLAG

The German National Library lists this publication in the German National Bibliography; detailed bibliographic information is available online at http://www.dnb.de.

Ch. Links Verlag is an imprint of Aufbau Verlage GmbH & Co. KG

The first edition was published in German by Ch. Links Verlag GmbH, 2019
www.aufbau-verlage.de/ch-links-verlag
Prinzenstrasse 85, 10969 Berlin
Cover design: Ch. Links Verlag, using a photo of the panda Meng Meng at Zoo Berlin (Steffen Freiling)
Typesetting: Nadja Caspar, Ch. Links Verlag
Printing and binding: Graspo CZ, Zlín, Czech Republic

ISBN 978-3-96289-221-0

Contents

Introduction: A tale of power plays and harmonious habitats

The "Zoological Garden by Berlin" opened on 1 August 1844 as the first modern zoo in Germany. Although the Zoo owed its existence to the favour of the Prussian king Friedrich Wilhelm IV, from the outset it was also a project of the citizens of Berlin, who gladly supported "their" zoological garden as shareholders and visitors. At the time of its opening, Zoologischer Garten Berlin was the ninth-oldest zoo in the world, and over the following decades it became the city's most popular attraction and a centre of social and recreational life.

However, the Zoo was also utilised as a tool for colonialist propaganda and aggrandisement. The Prussian court used the Zoo to ostentatiously display its might, and for a time it even exhibited human beings from outside Europe in so-called "ethnological exhibitions". During the Third Reich, the Zoo's Jewish shareholders were ousted and the workforce was Nazified. At the end of World War II, the German capital's zoological garden lay in ruins. Today, more than three million people from Germany and abroad visit Zoo and Aquarium Berlin each year.

This book tells the long and not always illustrious story of Berlin's Zoological Garden from its humble beginnings to the present day. A zoo is, of course, much more than a place of education and amusement; it is where encounters occur between humans and animals. For many people, it is the only place they get the opportunity to look such animals directly in the eye. The way in which zoos design those encounters always says a lot about societal attitudes towards animals and our relationship with them. And so the history of a zoo is a history of humans, as well.

Most books about zoos give a linear representation of the institution's successes, detailing developments in its keeping of and knowledge about animals. They begin perhaps by talking about royal menageries as the antithesis of a modern zoo, and finish by portraying their own zoo as a kind of Noah's Ark for at-risk species. But actually, conditions were not necessarily all that bad in the royal menageries – it was rather the systematic zoos of the nineteenth century that were inclined to squeeze their animals into small enclosures. And even the bar-free facilities first popularised by Carl Hagenbeck in the early twentieth century were not necessarily particularly animal-friendly if they were not of adequate size. The fact that zoo directors' ideas about

Western lowland gorilla Sango, 2019

Gorilla

The gorilla is the largest of the great apes. These impressive primates inhabit the rainforests and mountain forests of equatorial Africa, where they live in groups of up to forty members led by the strongest male. Adult males are known as "silverbacks" for the silver-grey hair they develop on their back and rump. Gorilla mothers have a gestation period of around nine months, after which they suckle their young for three and a half years. That means mother and young stay together for a good long while.

Gorillas communicate using hums, grunts and barks. Their diet consists mainly of roots, shoots, leaves, bark and tubers. The diurnal apes spend most of their time on the ground. In the evening, each gorilla builds its own sleeping nest. Gorillas are under severe threat in their jungle home due to the exploitation of natural resources and ongoing military conflicts. Like all great apes, gorillas have a special impact on zoo visitors. Their penetrating, intelligent eyes blur the line between human and animal. People are also fascinated by their obvious physical might, although gorillas are actually very gentle by nature.

Western lowland gorilla Djambala, 2013

Western lowland gorilla Sango, 2019

The first gorilla brought to Europe alive arrived in Berlin in 1876. The male, known by the name M'Pungu, moved into the aquarium on Unter den Linden that had been built by renowned zoologist Alfred Brehm. Unfortunately, M'Pungu only survived one year. By the beginning of the twentieth century, new young gorillas were being brought to Europe year after year, in many cases only to die soon after their arrival. At that time, not enough was known about how to properly look after the apes. Everyone involved in the business fully accepted that an entire gorilla group would have to be killed in order to capture a single juvenile. In the spring of 1928, young male Bobby arrived at the Zoo in Berlin. Unlike so many of his fellows before him, Bobby received good care and became the first gorilla to reach adulthood in a zoo. When Bobby died of appendicitis on 1 August 1935, he weighed an impressive 262 kilos. He is eternalised at the Zoo as a granite sculpture, which countless children have posed next to for photographs over the years, as well as in Zoo Berlin's logo.

animal-keeping reflected general attitudes towards animals among society may seem obvious, but it is important to bear that in mind.

The history of Zoo Berlin is also a history of the German capital and thus of Germany, too. This book explores the Zoo's origins in the Kingdom of Prussia and its history throughout the German Empire, the first German democracy from 1918 to 1933, the Third Reich, divided Germany, and finally its present-day status as one of two zoos in the capital of the reunited Federal Republic. A timeline at the bottom of most pages illustrates the historical development, presenting the most important events in the history of Berlin and Germany. The founding of the Empire, the two world wars, democracy, Nazism, the Cold War and German Unity all shaped the destiny of Berlin's zoological garden.

A history of the Zoo would be incomplete if it didn't also make frequent mention of the Tierpark – "animal park" – in the eastern part of the city, founded in 1955. First a Cold War rival and then a partner, the Tierpark has always influenced developments at Zoo Berlin. Tierpark Berlin-Friedrichsfelde GmbH has been a subsidiary of Zoologischer Garten Berlin AG since 1993, and since 2014 they have had a shared management board and managing director. And so in cases where the paths of the two zoos intersect or where they have influenced one another, this book looks at both Zoo and Tierpark Berlin. That is also true of the days when Stasi spies were doing everything they could to protect the reputation of "their" zoo.

However, a book that deals exclusively with the Tierpark's history is yet to be written. This publication is primarily about Zoo Berlin, the oldest existing zoological facility in Germany. The Aquarium is included as part of the Zoo.

Of course, the history of any zoo also has to be about its animals. Zoo Berlin's main characters would fade into obscurity if historical documents and old photographs were the only things to be shown. Therefore, at the beginning of each chapter, the book introduces an animal species that lives at Zoo or Aquarium Berlin. The various stages in the Zoological Garden's history can often be illustrated via those animals. For example, when was the species first displayed at Zoo Berlin? What were the pedagogical and political focuses of the Zoo directors? Each species represents an era of zoological history in Berlin while also demonstrating the diversity of species at the Zoo and Aquarium.

Some readers who already know a lot about Zoo Berlin might wonder why certain information seems to be missing: why is there no mention of this or that building, of this or that animal? Isn't the exhibition of animals that are now extinct – such as the quagga in the nineteenth century – or the first successful breeding of okapis or black rhinos more important than some seemingly trivial anecdote? From a zoological perspective, breeding successes are without doubt highly interesting, but this book is not intended to provide a mere list of zoological facts. Stories such as the one about the ingenuity of Berlin's only female zoo director to date, Katharina Heinroth, in overcoming the feed shortage during the Berlin Blockade are more exciting aspects of the Zoo's history than a full list of all Berlin's elephant births, for example.

But any history of Zoo Berlin must certainly mention its many animal celebrities over the years: gorilla Bobby, giraffe Rieke, chimpanzee Jonny, legendary hippo bull Knautschke, Berlin's first panda Bao Bao, global superstar Knut the polar bear. All these animals were beloved by Berliners in their day, and have thus been given plenty of space on the following pages.

This book was facilitated and generously supported by the supervisory board and management board of Zoologischer Garten AG. The Zoo archives were assembled by archivist Raphael Hartisch, but countless employees of Zoo, Aquarium and Tierpark Berlin also volunteered their advice and assistance, their knowledge and their memories. The IT department secured and reorganised the archives, transferred files, and helped set up a research infrastructure. The Zoo's curators and veterinarians patiently answered every question and provided insight into their work, while the administrators gladly integrated an external documentation centre with its inquisitive historian into their daily working life. Last but not least, the team of animal keepers helped me out a great deal by showing me around and introducing me to the animals in their care. This magnificent team of dedicated keepers may not always be visible to the public, but they are always hard at work behind the scenes taking care of the Zoo's and this book's leading protagonists – the animals!

Adolf Etzner, bird's-eye view of the Zoological Garden of Berlin, 1873

Rulers and their animals: The predecessors to Zoo Berlin

Animals as status symbols

Keeping "wild" animals in captivity is an ancient practice. Even during prehistoric times, humans would keep animals for their meat or as totems in sacred places. But it was not until people began collecting multiple animals of various species for the primary purpose of displaying them that we can really speak of predecessors of modern-day zoos. There is evidence that some early advanced civilisations kept animals for display purposes.[1] Findings by Egyptologists suggest that the courts of the pharaohs housed animal collections from the second millennium BCE onwards. Chinese king Wen Wang created a "Garden of Knowledge" in around 1150 BCE that contained a wide array of animals. The Book of Kings in the Old Testament suggests that King Solomon kept wild animals, and in the Mesopotamian kingdoms of Sumeria, Babylon and Assyria, animals were exhibited in specially designed gardens. Rulers received animals as gifts from other powerful leaders or collected them on their own initiative to impress their subjects and guests. Such animals were status symbols intended to signify the pre-eminence of mighty rulers who had subjugated foreign lands and even nature itself.

Animal-keeping played a more minor role in Ancient Greece and Ancient Rome, where people kept animals either to eat them or for entertainment purposes, pitting them against one another in elaborate circuses, or even setting them on unfortunate humans. Roman emperors would also capture the animals of vanquished foes in order to demonstrate their own grandeur to their subjects back home. It was the rulers of the Ptolemaic Kingdom who created the first real predecessor to modern zoos – in Alexandria, which was the capital of Egypt during the Hellenistic period.

In the Middle Ages, captive animals became symbols of power. The first Holy Roman Emperor, Charlemagne, King of the Franks, maintained menageries of many exotic animals at several of his royal courts. Often, these animals were diplomatic gifts from other rulers, such as the elephant presented to Charlemagne by the Abbasid caliph Harun al-Rashid. Emperor Friedrich II of the Hohenstaufen dynasty kept animals in parks in order to study them. In the late Middle Ages, provincial princes increasingly began to estab-

An Edwards's pheasant, 2018

Common pheasant

The common pheasant (*Phasianus colchicus*) belongs to the order Galliformes, which includes chickens and other landfowl. The males stand out with their colourful plumage and long tail feathers, while the females are usually more of a speckled brown, which keeps them well camouflaged. These pheasants play a central role in the myths and symbolism of their original homeland – a wide area of land stretching from China to the Caucasus.

Common pheasants are usually found in semi-open terrain, sparse forests with undergrowth, or wetlands with plenty of reeds. These habitats provide cover as well as open areas for foraging. The birds feed mainly on vegetation but also eat insects and other small animals.

Wild pheasants have long been popular game birds, famed for their tasty flesh. The Romans kept them as livestock and spread them throughout Central Europe. From the Middle Ages, the birds were kept in royal courts and monasteries in "pheasantries". Throughout history, pheasant feathers have been a popular fashion accessory in many cultures, adorning hats, helmets, uniforms, ceremonial attire and fans.

Many ramblers encounter pheasants during the birds' mating season, when the males mark their territories with characteristic calls and wing-beating. They mate with multiple passing hens, who then stay in the area to hatch their eggs.

The common pheasant has always had close associations with the Zoo in Berlin, which had its own pheasantry from the very beginning. The Zoo actually started life on the grounds of the royal pheasantry in the Tiergarten park. Today, it is particularly committed to breeding Vietnamese pheasants – a species that is found wild only in Vietnam and was almost wiped out in the Vietnam War. These birds were particularly hard hit by the devastation of their habitat arising from the liberal use of the defoliant Agent Orange by US forces. Hopes are high that the Berlin-managed studbook will strengthen the population of the species so that successful reintroduction programmes can begin.

A Lady Amherst's pheasant

Interior view of the Pheasantry

lish their own animal collections in a bid not to be outdone by their more powerful counterparts. This practice led to the establishment of numerous collections of exotic and native animals in the fifteenth and sixteenth centuries – in Stuttgart and Frankfurt am Main, for example, at the court of the Medici family in Florence, and at the Fuggers' premises in Augsburg.

From royal menageries to zoos

Naturally, Louis XIV of France – the famous "Sun King" – could not do without his own collection of exotic animals. A menagerie was built at Versailles in 1662, the first major project in the park. This Royal Menagerie of Versailles was a series of animal enclosures fanning out from a central octagonal building that afforded the king a panoramic view of his collection. From this vantage point, he could proudly overlook a natural world that he had structured according to his will. Of course, only the king, his courtiers and honoured guests were permitted to see these miracles of creation. In 1793, during the French Revolution, the surviving animals were transferred to the Parisian botanical garden, the Jardin des Plantes.

Meanwhile, in around 1750, driven in equal measure by curiosity and the desire for prestige, Holy Roman Emperor Franz I decided to establish a collection of exotic animals at his Schönbrunn Palace in Vienna, based on the Royal Menagerie of Versailles. The menagerie in Schönbrunn was a circular system of enclosures radiating out from a central belvedere. In 1778, under the reign of Empress Maria Theresa, the collection was opened to the public. Today, Tiergarten Schönbrunn is the world's oldest zoo still in operation.

Jean Nicolas Jadot de Ville-Issey, design drawings for the Schönbrunn menagerie, circa 1751

For a long time, the only way ordinary people could see exotic wild animals was to visit so-called travelling menageries. These would go on tour from place to place, exhibiting their animals for a few weeks at a time. The scant descriptions and explanations of the animals on display were born more out of the showmen's imagination than any actual scientific knowledge.

Pheasantries and menageries

All the rulers of Brandenburg, from the Ascanians to the Hohenzollerns, probably kept wild animals in some way or another. In 1742, King

Friedrich II ("Frederick the Great" of Prussia) had a pheasantry built in the southwestern part of the former royal hunting ground just outside Berlin, which had been transformed into a pleasure garden known as the "Grosser Tiergarten". From 1794 to 1797, his nephew and successor Friedrich Wilhelm II established his summer residence on "Peacock Island" (*Pfaueninsel*) in the River Havel southwest of the city. From 1816, *his* son Friedrich Wilhelm III had the parkland on Peacock Island redesigned by landscape architect Peter Joseph Lenné, with auxiliary buildings planned by renowned architects Karl Friedrich Schinkel and Albert Dietrich Schadow. The king began to keep exotic animals on the island from around 1820. The number of animals living there fluctuated, but the collection appears to have been anything but small.

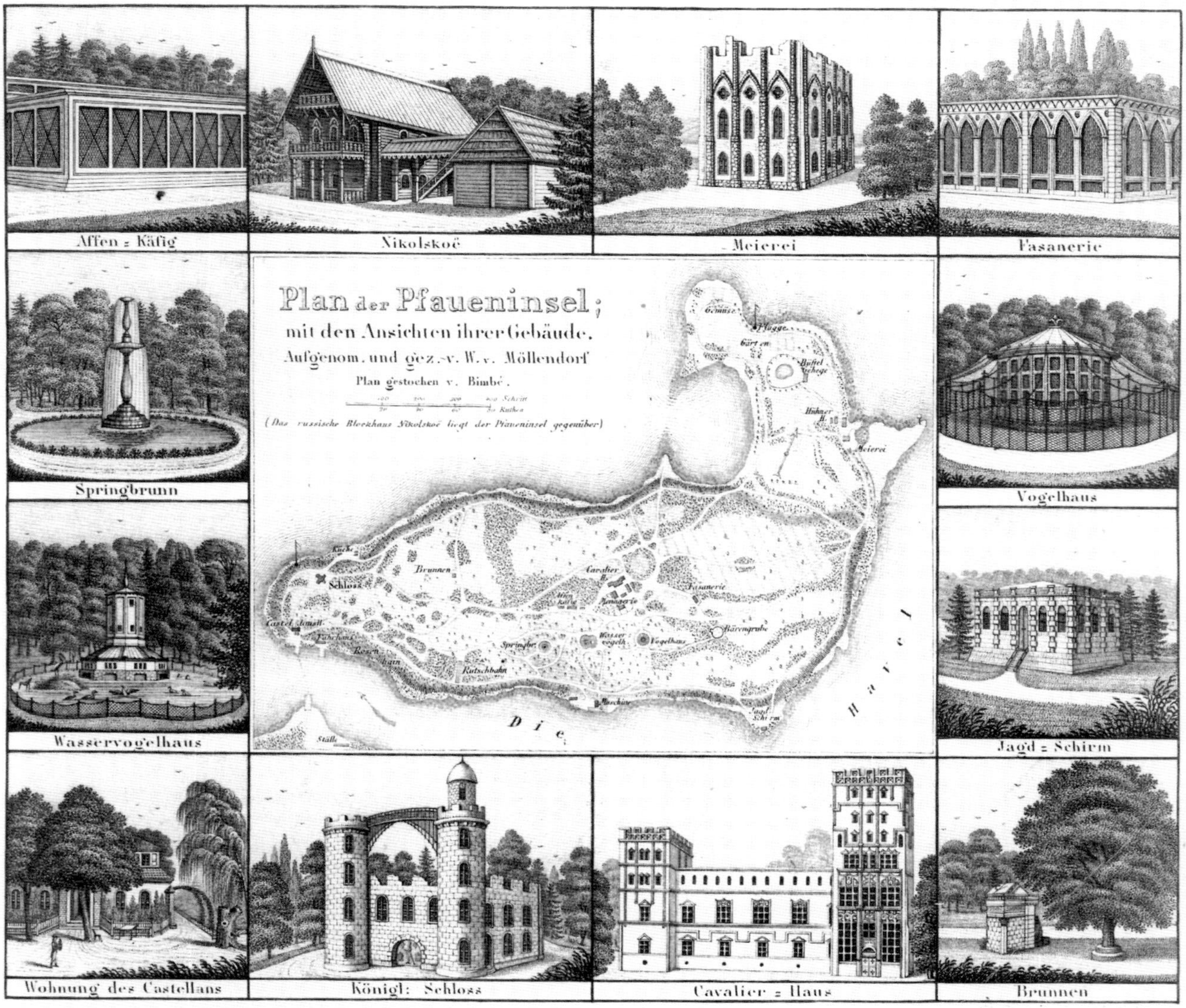

Bimbé and Ludwig Meyer after Wilhelm von Möllendorf, Peacock Island, map and illustrations, 1827

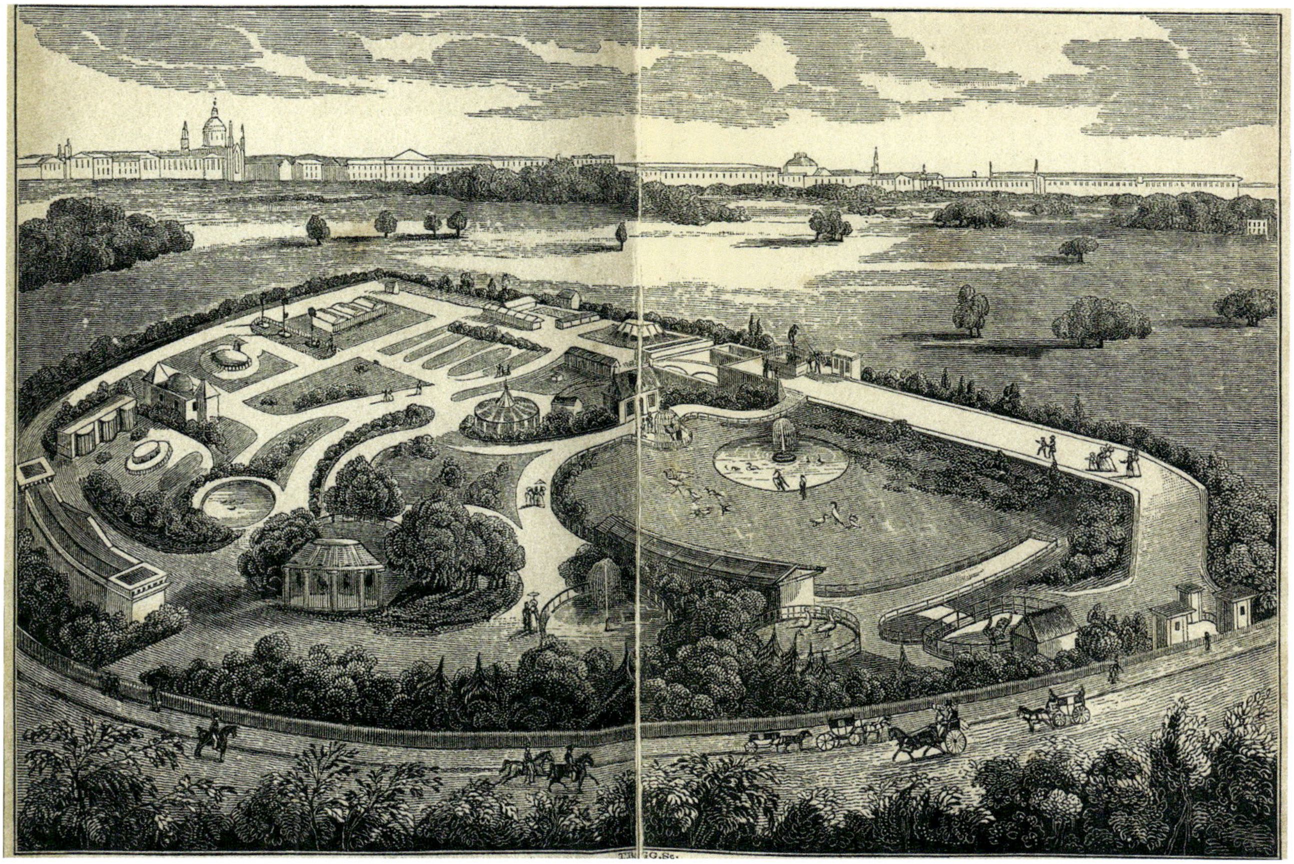

Lithograph of Regent's Park Zoo, 1830

In 1832, almost 850 animals were housed on Peacock Island. In addition to many pheasants, peacocks and native birds, they included more exotic animals such as kangaroos, several species of monkey, various bears, and even a lion.[2] Although the primary purpose of these animals was to delight and amuse the king, on certain days the menagerie was open to the public. Physician and zoologist Martin Hinrich Carl Lichtenstein oversaw the collection. He had been appointed the first professor of zoology at the newly founded Friedrich-Wilhelms-Universität in Berlin in 1811 and became director of the Zoological Museum in 1813. A real man of science, Lichtenstein had ambitious plans for the menagerie's collection.

Hinrich Lichtenstein (1780–1857), circa 1850

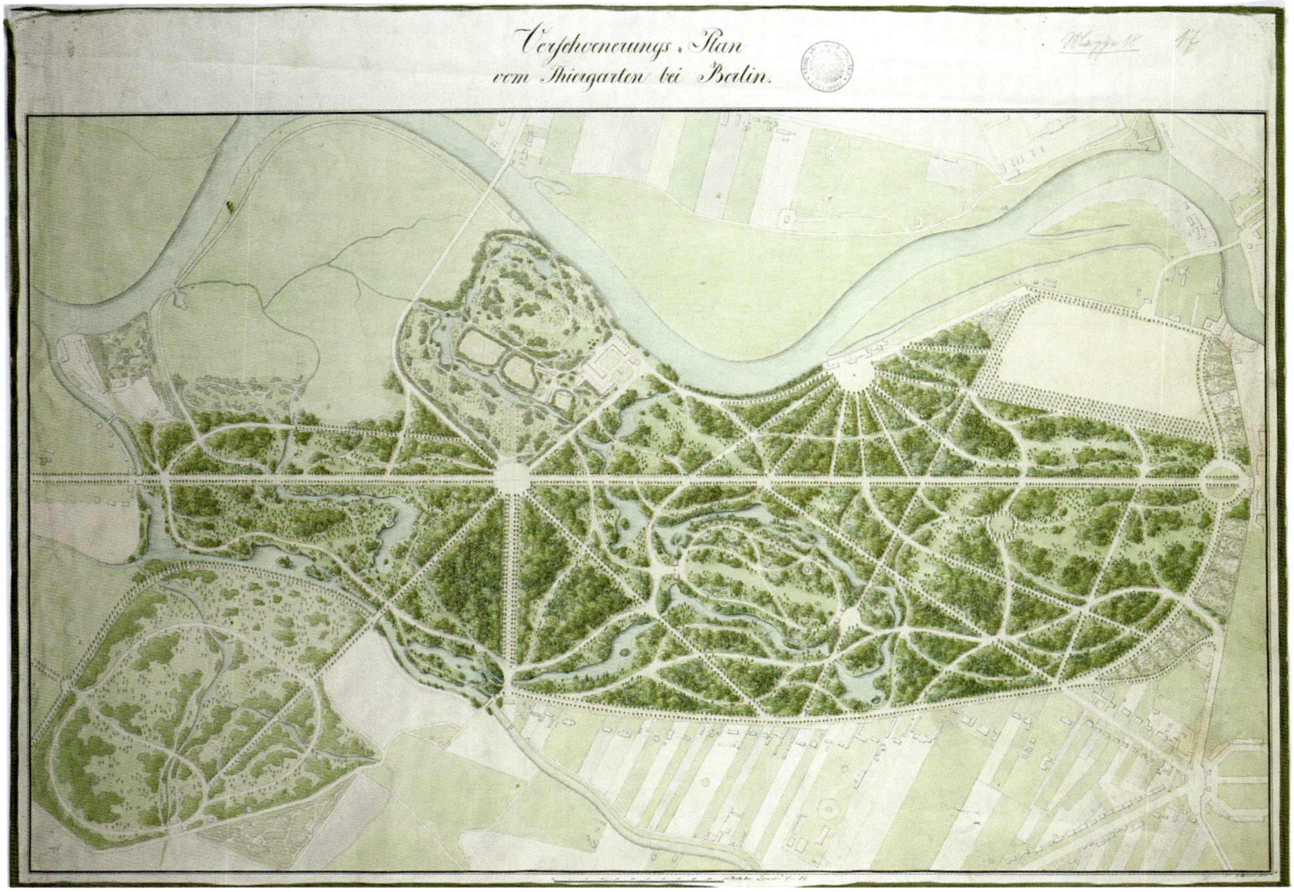

Peter Joseph Lenné, plans to improve the Tiergarten royal hunting park by Berlin from 1832. An island surrounded by watercourses can be seen south of the lateral axis, complete with paths and enclosures.

London sets an example

The Zoological Society of London was founded in 1826. Two years later, the society opened a zoo in a section of Regent's Park that it had leased from the English Crown for a nominal fee. At first, only members of the society were granted admission to the zoo, but from 1847 it opened to the public as well.

Lichtenstein dreamed of establishing a similar "scientific zoo" in Berlin, a zoo with a primary focus on research and education for the benefit of the state and society. With support from his friend, renowned naturalist and explorer Alexander von Humboldt, in the summer of 1833 he approached Friedrich Wilhelm III with a special request: he asked the king to finance an official study trip to London so he could gather important insights into animal-keeping that could benefit the Peacock Island menagerie. The king agreed, and so Lichtenstein journeyed to England in autumn 1833. He returned from the trip more convinced than ever that Berlin should have its own zoo.[3] And he already had the perfect location in mind: the Grosser Tiergarten. He also had the committed support of Peter Joseph Lenné, now director general of the royal Prussian palaces and parks, who had already incorporated a possible zoo into his future plans for the park.

This zoo was to be created on an area of about ten hectares in the southern part of the Tiergarten, which would have made it bigger than its London counterpart. London Zoo's location at the heart of a royal park was probably the inspiration for Lenné and Lichtenstein.[4] However, nothing could happen without the king's permission – and this was not granted. The monarch's passion for the project had apparently faded. The ambitious plans dreamed up by Lichtenstein, Humboldt and Lenné had to wait for a subsequent ruler to display an interest.

The hopes of all zoo enthusiasts were therefore pinned on the next king, Friedrich Wilhelm IV, who ascended the throne in 1840. That same summer, Lichtenstein retired to Cunersdorf manor in Brandenburg, where he wrote a memorandum with the goal of persuading the king to establish a zoological garden near Berlin. In the autumn, Alexander von Humboldt presented this memorandum to Friedrich Wilhelm IV. It piqued the monarch's interest, and on 31 January 1841 he issued an order for Lichtenstein and Humboldt to form a committee with Adalbert von Ladenberg of the Prussian Ministry for Intellectual and Educational Affairs and Director General Lenné with the purpose of establishing a zoo.[5]

Paul Sigmund Habelmann after Emma Gaggiotti-Richards, Alexander von Humboldt, circa 1850

A zoological garden by Berlin, 1844–1870

Makeshift beginnings

The idea of a place where large numbers of people could admire and study animals was very much in keeping with the zeitgeist of mid-nineteenth-century Europe. Learning, enlightenment and moral education were bourgeois ideals as well as goals of the Prussian state. Also, Friedrich Wilhelm IV knew that a zoo would make his royal city even more appealing. In May 1841, the zoo committee assembled at the Grosser Tiergarten's pheasantry, close to the Brandenburg Gate. The committee members inspected the site and decided to propose to the king that the grounds and some of its structures should become part of a "scientific institute protected and supervised by the state".[1] In return, the zoo would pass all animal carcasses on to the royal collections and grant the court a right of first refusal on animal sales. Scholars should also be granted access to the zoo for research purposes. Following the recommendations of the committee, on 8 September 1841 the king granted a "cabinet order" giving his consent to the establishment of a zoological garden on the site of the royal pheasantry. In that order, the king granted the future zoo use of the grounds, buildings and fences without charge and promised a loan of 15,000 thalers as well as gifts of animals and cages from his menagerie on Peacock Island.[2]

Landscaping work began in July 1842, and in the spring of 1844 the first animals from Peacock Island arrived in Berlin. The monarch's initial delivery comprised a total of forty-seven animals, with others available in principle. Friedrich Wilhelm had already provided start-up capital in the form of a five-year interest-free loan of 15,000 thalers. Shortly afterwards, however, this sum had to be augmented with a further 10,000 thalers "to cover the costs for the construction of the monkey house, the llama house, the cage for large predators, as well as small fences, benches and the like".[3] In 1844, the new zoo was finally ready to welcome visitors.

Lichtenstein opened the "Zoological Garden by Berlin" on 1 August 1844 – to relatively little fanfare. Lichtenstein was the only one of the four committee members present. Friedrich Wilhelm was also absent. A few days earlier, there had been an attempt on his life and he was thus avoiding public appearances.

A map of the Zoo's layout from 1845 shows its humble beginnings in that first year. Only a

Elephant bull Victor, 2018

Asian elephant

Elephants are unquestionably one of the most popular zoo animals. There are three extant elephant species: the African bush elephant (*Loxodonta africana*), the slightly smaller African forest elephant (*Loxodonta cyclotis*), and the Asian elephant (*Elephas maximus*). All three species have been kept at different times at Zoo Berlin, sometimes in the same habitat. In the mid-1990s, the last African elephant at the Zoo – a female named Lilak – moved to the Tierpark. Since then, Zoo Berlin has been home to Asian elephants only.

Elephants are the largest land mammals on Earth. They also have the longest gestation period of any animal: a female elephant carries her young for up to 22 months. It's no surprise then that elephants have the highest birth weight of all land mammals, at around 100 kilos. But the elephant is a record-holders in other areas, too. For example, its trunk contains more than 40,000 muscles – the most found in any organ.

Anchali between the legs of her mother Pang Pha, 2012

In Asian elephants, tusks are usually only visible on the males, whereas the females of both African species also develop the elongated incisors. In addition to their tusks, elephants have a set of about four non-visible molars that are continually replaced throughout their life. These teeth actually determine the lifespan of an elephant: once the sixth and final set has been worn down, the animal struggles to eat. This usually occurs between the ages of 45 and 50. With good care, zoo elephants can live up to 60 years.

Female elephants Anchali and Pang Pha splashing in their pool, 2019

The natural range of Asian elephants originally comprised a large swathe of land across all of Southeast Asia and the adjacent islands. Today, all three elephant species are endangered due to loss of their natural habitat. They are also still regularly killed by poachers for their ivory tusks.

The first elephant to come to Berlin was a male Asian elephant named Boy who arrived in 1857, thirteen years after the opening of the Zoological Garden in Berlin. In recent decades, most of Berlin's elephant acquisitions have been organised exchanges with other zoos as part of international breeding programmes.

few enclosures and animal houses dot the landscaped garden created by Peter Joseph Lenné on the site of the royal pheasantry. At that time, the Zoo's entrance was located on the northeastern side of the grounds facing the Tiergarten, beyond which lay the city of Berlin. On this south-facing map, the entrance can be seen on the bottom left.

were native animals. The most exotic inhabitants were a leopard, several camels, and a small group of kangaroos.[4] The majority of these originated from the menagerie on Peacock Island, which was now closed. The modest collection came in for a great deal of criticism, especially since the admission fee of five *Silbergroschen* (a Prussian silver coin worth 60 pfennigs) per adult was not

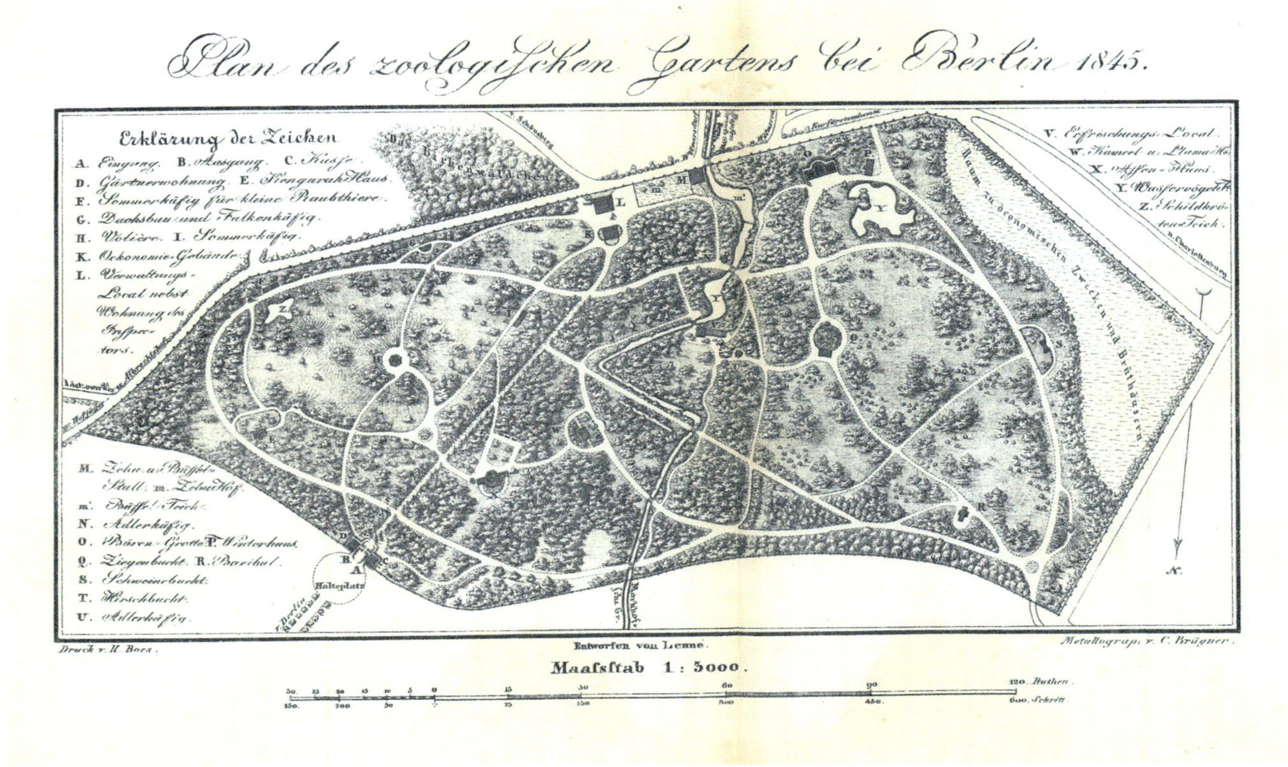

Peter Joseph Lenné, map of the Zoological Garden, 1845

At the time of its opening, the Zoological Garden was not yet complete and the number of animals on display was rather small in relation to the size of the grounds. For decades, this 22-hectare park remained the largest of its kind in the world in terms of area. The printed guide to the Zoo published in June 1845 included fewer than a hundred species, most of which

exactly cheap. Only years later did the Zoo introduce "low-cost Sundays", when admission was half the price.

Early visitors also bemoaned the Zoo's lack of refreshments. In an open letter to the newspapers, Lichtenstein asked for the public's patience, explaining that a restaurant was under construction. He also announced that smoking was per-

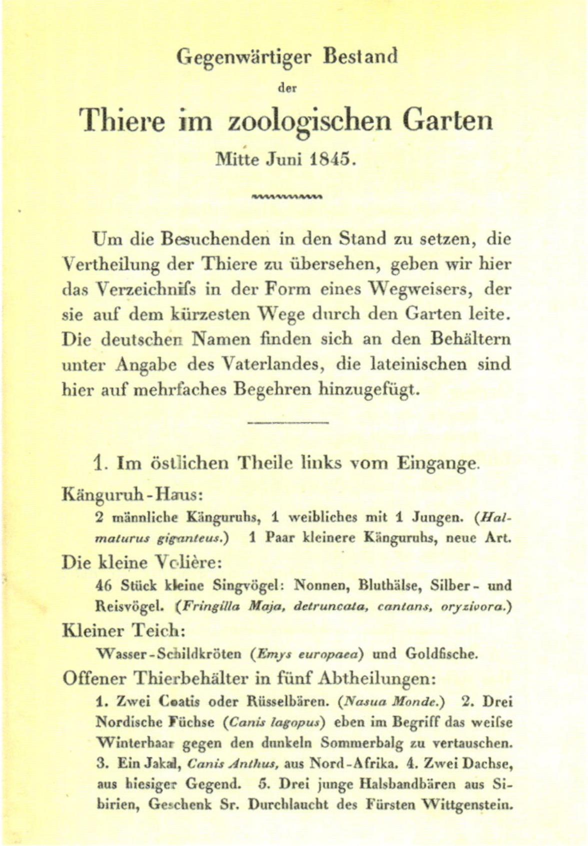

Gegenwärtiger Bestand

der

Thiere im zoologischen Garten

Mitte Juni 1845.

Um die Besuchenden in den Stand zu setzen, die Vertheilung der Thiere zu übersehen, geben wir hier das Verzeichniſs in der Form eines Wegweisers, der sie auf dem kürzesten Wege durch den Garten leite. Die deutschen Namen finden sich an den Behältern unter Angabe des Vaterlandes, die lateinischen sind hier auf mehrfaches Begehren hinzugefügt.

1. Im östlichen Theile links vom Eingange.

Känguruh-Haus:
2 männliche Känguruhs, 1 weibliches mit 1 Jungen. (*Halmaturus giganteus.*) 1 Paar kleinere Känguruhs, neue Art.

Die kleine Volière:
46 Stück kleine Singvögel: Nonnen, Bluthälse, Silber- und Reisvögel. (*Fringilla Maja, detruncata, cantans, oryzivora.*)

Kleiner Teich:
Wasser-Schildkröten (*Emys europaea*) und Goldfische.

Offener Thierbehälter in fünf Abtheilungen:
1. Zwei Coatis oder Rüsselbären. (*Nasua Monde.*) 2. Drei Nordische Füchse (*Canis lagopus*) eben im Begriff das weiſse Winterhaar gegen den dunkeln Sommerbalg zu vertauschen. 3. Ein Jakal, *Canis Anthus*, aus Nord-Afrika. 4. Zwei Dachse, aus hiesiger Gegend. 5. Drei junge Halsbandbären aus Sibirien, Geschenk Sr. Durchlaucht des Fürsten Wittgenstein.

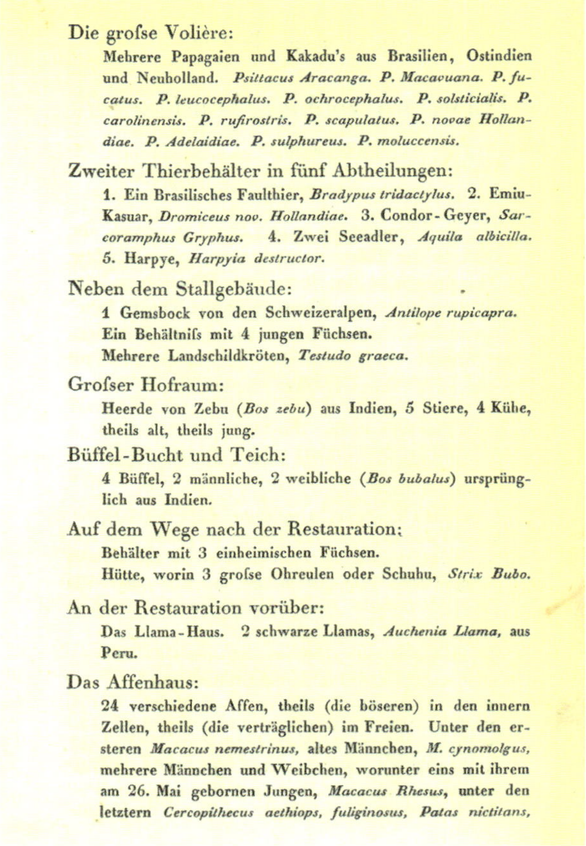

Die groſse Volière:
Mehrere Papagaien und Kakadu's aus Brasilien, Ostindien und Neuholland. *Psittacus Aracanga. P. Macavuana. P. fucatus. P. leucocephalus. P. ochrocephalus. P. solsticialis. P. carolinensis. P. rufirostris. P. scapulatus. P. novae Hollandiae. P. Adelaidiae. P. sulphureus. P. moluccensis.*

Zweiter Thierbehälter in fünf Abtheilungen:
1. Ein Brasilisches Faulthier, *Bradypus tridactylus.* 2. Emiu-Kasuar, *Dromiceus nov. Hollandiae.* 3. Condor-Geyer, *Sarcoramphus Gryphus.* 4. Zwei Seeadler, *Aquila albicilla.* 5. Harpye, *Harpyia destructor.*

Neben dem Stallgebäude:
1 Gemsbock von den Schweizeralpen, *Antilope rupicapra.*
Ein Behältniſs mit 4 jungen Füchsen.
Mehrere Landschildkröten, *Testudo graeca.*

Groſser Hofraum:
Heerde von Zebu (*Bos zebu*) aus Indien, 5 Stiere, 4 Kühe, theils alt, theils jung.

Büffel-Bucht und Teich:
4 Büffel, 2 männliche, 2 weibliche (*Bos bubalus*) ursprünglich aus Indien.

Auf dem Wege nach der Restauration:
Behälter mit 3 einheimischen Füchsen.
Hütte, worin 3 groſse Ohreulen oder Schuhu, *Strix Bubo.*

An der Restauration vorüber:
Das Llama-Haus. 2 schwarze Llamas, *Auchenia Llama,* aus Peru.

Das Affenhaus:
24 verschiedene Affen, theils (die böseren) in den innern Zellen, theils (die verträglichen) im Freien. Unter den ersteren *Macacus nemestrinus,* altes Männchen, *M. cynomolgus,* mehrere Männchen und Weibchen, worunter eins mit ihrem am 26. Mai gebornen Jungen, *Macacus Rhesus,* unter den letztern *Cercopithecus aethiops, fuliginosus, Patas nictitans,*

Animal inventory of the Zoological Garden, 1845

mitted on the designated paths as long as sufficient distance was maintained from the cages. This was at a time when smoking was strictly prohibited on the streets of Berlin – a ban that would be lifted in 1848.

Above all, however, Lichtenstein asked for visitors' understanding: "I must therefore reiterate that this is a public facility that is still in the making and will only achieve its full brilliance at a later date, once it has received the necessary encouragement and support. [...] We ask that the animals present here at the current time be regarded solely as an embellishment to the grounds and that, for now, visitors try to overlook the fact that building materials are still piled up in many places, that old wooden buildings have not yet been removed, and that new buildings are still under construction [...]."[5]

For Lichtenstein, the purpose of the Zoo was to combine "rigorous scientific study" with "instructive entertainment".[6]

In its first year, the Zoo welcomed several hundred visitors on some days, while on other days there were none at all. The general lack of visitor interest and the significantly higher costs than originally forecast meant that the Zoo quickly found itself in major financial difficulties. The income from the first year amounted to a dismal 4,700 thalers. The following year it rose to 9,300 thalers, and in 1846 it increased further to 14,000. But in 1847 the Zoo's income dropped by half.[7] People found the makeshift nature of

June 1844
Silesian weavers' uprising

26 July 1844
Assassination attempt on Friedrich Wilhelm IV

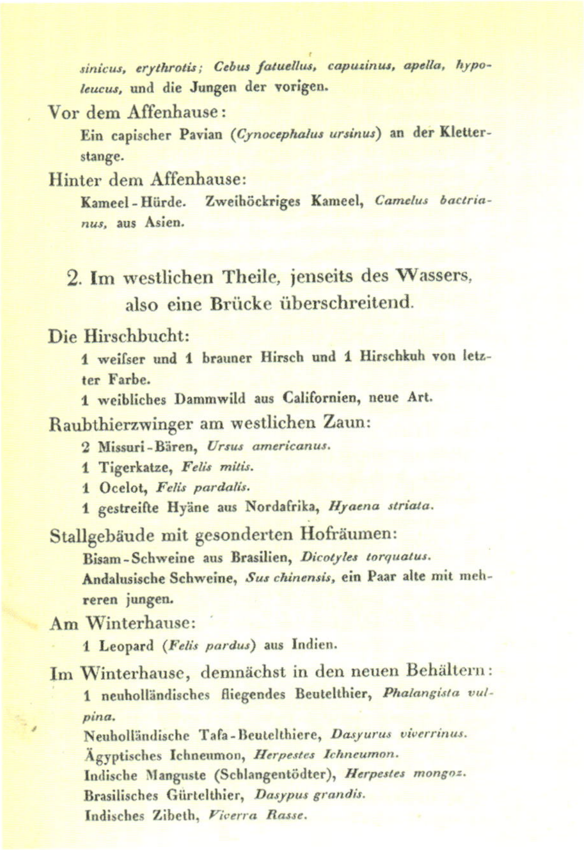

sinicus, erythrotis; Cebus fatuellus, capuzinus, apella, hypoleucus, und die Jungen der vorigen.

Vor dem Affenhause:

Ein capischer Pavian (*Cynocephalus ursinus*) an der Kletterstange.

Hinter dem Affenhause:

Kameel-Hürde. Zweihöckriges Kameel, *Camelus bactrianus,* aus Asien.

2. Im westlichen Theile, jenseits des Wassers, also eine Brücke überschreitend.

Die Hirschbucht:

1 weifser und 1 brauner Hirsch und 1 Hirschkuh von letzter Farbe.
1 weibliches Dammwild aus Californien, neue Art.

Raubthierzwinger am westlichen Zaun:

2 Missuri-Bären, *Ursus americanus.*
1 Tigerkatze, *Felis mitis.*
1 Ocelot, *Felis pardalis.*
1 gestreifte Hyäne aus Nordafrika, *Hyaena striata.*

Stallgebäude mit gesonderten Hofräumen:

Bisam-Schweine aus Brasilien, *Dicotyles torquatus.*
Andalusische Schweine, *Sus chinensis,* ein Paar alte mit mehreren jungen.

Am Winterhause:

1 Leopard (*Felis pardus*) aus Indien.

Im Winterhause, demnächst in den neuen Behältern:

1 neuholländisches fliegendes Beutelthier, *Phalangista vulpina.*
Neuholländische Tafa-Beutelthiere, *Dasyurus viverrinus.*
Ägyptisches Ichneumon, *Herpestes Ichneumon.*
Indische Manguste (Schlangentödter), *Herpestes mongoz.*
Brasilisches Gürtelthier, *Dasypus grandis.*
Indisches Zibeth, *Viverra Rasse.*

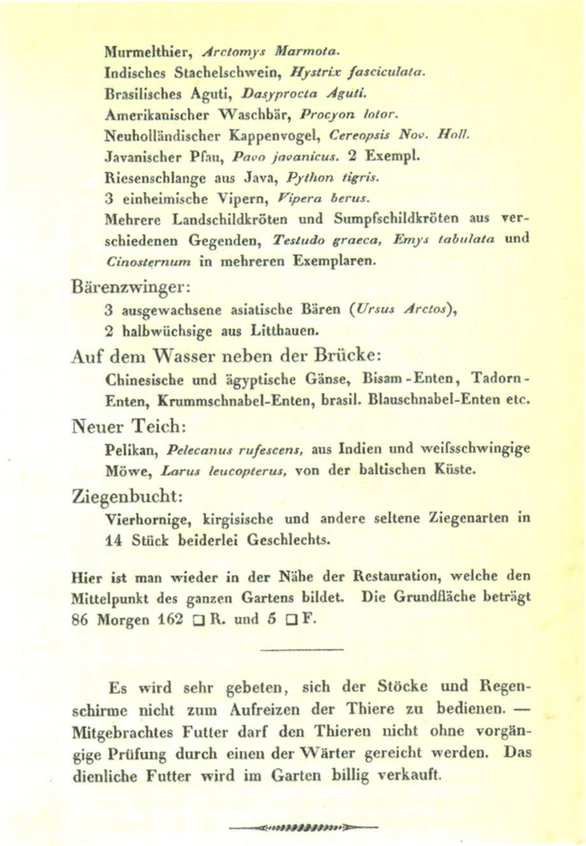

Murmelthier, *Arctomys Marmota.*
Indisches Stachelschwein, *Hystrix fasciculata.*
Brasilisches Aguti, *Dasyprocta Aguti.*
Amerikanischer Waschbär, *Procyon lotor.*
Neuholländischer Kappenvogel, *Cereopsis Nov. Holl.*
Javanischer Pfau, *Pavo javanicus.* 2 Exempl.
Riesenschlange aus Java, *Python tigris.*
3 einheimische Vipern, *Vipera berus.*
Mehrere Landschildkröten und Sumpfschildkröten aus verschiedenen Gegenden, *Testudo graeca, Emys tabulata* und *Cinosternum* in mehreren Exemplaren.

Bärenzwinger:

3 ausgewachsene asiatische Bären (*Ursus Arctos*),
2 halbwüchsige aus Litthauen.

Auf dem Wasser neben der Brücke:

Chinesische und ägyptische Gänse, Bisam-Enten, Tadorn-Enten, Krummschnabel-Enten, brasil. Blauschnabel-Enten etc.

Neuer Teich:

Pelikan, *Pelecanus rufescens,* aus Indien und weifsschwingige Möwe, *Larus leucopterus,* von der baltischen Küste.

Ziegenbucht:

Vierhornige, kirgisische und andere seltene Ziegenarten in 14 Stück beiderlei Geschlechts.

Hier ist man wieder in der Nähe der Restauration, welche den Mittelpunkt des ganzen Gartens bildet. Die Grundfläche beträgt 86 Morgen 162 □R. und 5 □F.

Es wird sehr gebeten, sich der Stöcke und Regenschirme nicht zum Aufreizen der Thiere zu bedienen. — Mitgebrachtes Futter darf den Thieren nicht ohne vorgängige Prüfung durch einen der Wärter gereicht werden. Das dienliche Futter wird im Garten billig verkauft.

the Zoo off-putting and only a few could afford the high admission prices. Around 60,000 people visited the Zoo in the first year of its existence, curious to see what it looked like – but that was mostly in the summer months. More than 13,000 visitors were recorded in July 1845 alone, and the Zoo's "Jubilee Sunday" on 3 August 1845 attracted almost 1,400 people on a single day. But those early visitors only got to see some 120 mammals, 100 birds, and 20 amphibians.[8]

The Zoo found itself in a vicious circle, as low income meant fewer exotic animals and fewer attractive animal houses. Well into the 1850s, some big cats were exhibited in open "receptacles" during the summer and had to be relocated to the few existing indoor spaces in the wintertime. The Zoo's managers were confronted with a constant dilemma: should they prioritise purchasing new and interesting animals in order to attract visitors, or should they first build the houses to put those animals in?

Informative signs were also few and far between in those early years. Instead, wardens posted by the entrance would tell visitors about what awaited them inside, and other wardens by the enclosures and cages would be on hand to provide additional information.[9] From April 1845, a guide was employed to take visitors around. That summer, the guide was granted a bonus so he could buy a new pair of boots, the old ones having worn out from the long distances covered on foot.[10] There is no way of knowing what anecdotes the staff were telling visitors at that time, but it is highly likely that some wardens embellished the truth in hopes of a bigger tip.

This regrettable situation was remedied to some degree by the printed guides that the Zoo subsequently began to publish. Seen from a modern perspective, the information in those early publications often appears superficial, even bizarre. Particularly striking is the fact that they

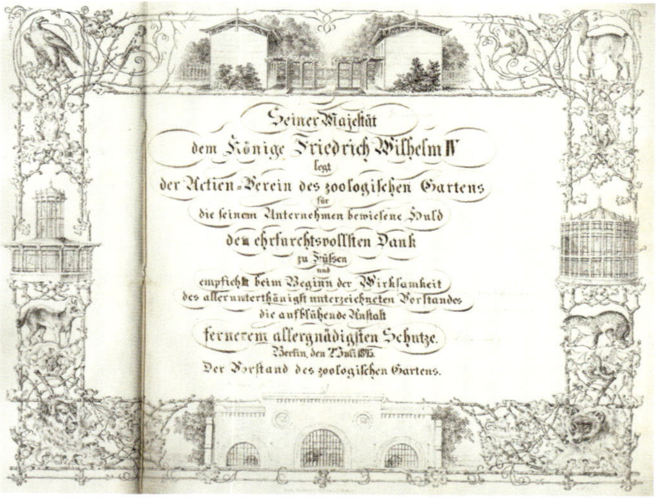
Seiner Majestät
dem Könige Friedrich Wilhelm IV
legt
der Actien-Verein des zoologischen Gartens
für
die seinem Unternehmen bewiesene Huld
den ehrfurchtsvollsten Dank
zu Füßen
und
empfiehlt beim Beginn der Wirksamkeit
des allerunterthänigst unterzeichneten Vorstandes
die aufblühende Anstalt
fernerem allergnädigsten Schutze.
Berlin, den 2[illegible] Juli 1845.
Der Vorstand des zoologischen Gartens.

Thank-you note from the management board to Friedrich Wilhelm IV, written on ornately headed share paper, 1845

focus mainly on the external characteristics of the animals and contain little information about their behaviour. That reflects the nature of zoological research in the mid-nineteenth century, which was still mainly concerned with laying down a systematised description of the animal kingdom rather than with exploring behavioural biology. In the early printed guides, the predator descriptions focus on the danger they represent to humans, and there is a strange emphasis on the might and ferocity of other animals, too. The 1849 guide asserts that condors kill bulls by tearing out their eyes and tongue, while the 1866 publication states that elephants live to be 200 years of age. The dissemination of "facts" like these shows to what degree Lichtenstein and his colleagues were dependent on hearsay and the descriptions provided by animal catchers, especially for the more exotic species.

Some early visitors had quite unrealistic wishes regarding the Zoo's opening hours. In April 1845, one visitor complained in an anonymous letter to the newspaper *Vossische Zeitung* that the Zoological Garden didn't open until eight o'clock in the morning despite the fact that on hot days a visit was only tolerable either very early or very late in the day. It should be remembered that at that time male visitors were still wearing top hats, stiff stand-up collars, and frock coats. Lichtenstein's reply appeared promptly in the same paper. He explained that the wardens had to muck out the enclosures before the Zoo could open. He also mentioned that certain gentlemen had previously been admitted before six o'clock in the morning, though on one warm night the nightwatchman had to turn away another group of gentlemen who were seeking admission between three and four o'clock![11] Generally, however, the keen interest displayed by a few was outweighed by a majority that was either indifferent or too poor to afford a ticket.

King Friedrich Wilhelm IV honoured the newly founded Zoological Garden with a visit one year after its opening. On the afternoon of 19 July 1845, he and the queen arrived from their summer residence in Charlottenburg and were received by the board of directors.[12] At that time, the Zoo had barely any major attractions to offer visitors. Many members of the public had already experienced exotic animals such as elephants, giraffes and monkeys in travelling menageries and as such expected to also encounter those animals

October 1846
Berlin's first horse-drawn omnibus lines go into operation

15 December 1846
The Berlin-Hamburg railway opens

1 January 1847
The first municipal gasworks open

at the Zoo. What they wanted more than anything was the chance to marvel at the "king of the beasts". At last, in September 1845, a lion from "Nubia" arrived. He was exhibited in the "Open Hall for Predators with Nine Sections" alongside a crabeater seal, a South American bush dog, a hyena, a leopard from India, a panther from Africa, an ocelot from Brazil, and prairie dogs from North America. There was also a special cage that displayed either a jaguar or a puma. Overall, that is quite an impressive collection of predators, but their numbers fluctuated greatly. The lion, for example, didn't survive long after his journey to Berlin, which would have taken months and been beset with hardships. At that time, there was insufficient knowledge of the conditions required for keeping such animals. After the lion's death, the Zoo again had to make do without one until the late summer of the following year, when a large shipment of animals arrived in Berlin after a long trip. The delivery included the extravagant gift of two lions, sent to the Prussian king from General Yusuf in French-occupied Algeria.[13]

In autumn 1846, the Zoo was able to announce that two mandrills and other monkeys had arrived, as well as a Brazilian tayra – a small predator in the weasel family. This was the first time a live specimen of the latter had been exhibited in a European animal collection. Though decidedly less exotic, the addition of four North Sea seals constituted a delightful spectacle for many visiting Berliners.[14]

In order to obtain urgently needed exotic animals at a low cost, in the summer of 1845 Lichtenstein decided that, in view of the Zoo's lack of funds, he would approach the royal Prussian consuls in the most important foreign trading and coastal cities with a request for animal donations.[15] The Zoo simply did not have the money to make any purchases that year. Although the Berlin Zoological Garden did round off the year 1846 with a scant profit of 561 thalers, it was forced to use all of that and more to purchase animal feed and pay off debts from the previous year.[16]

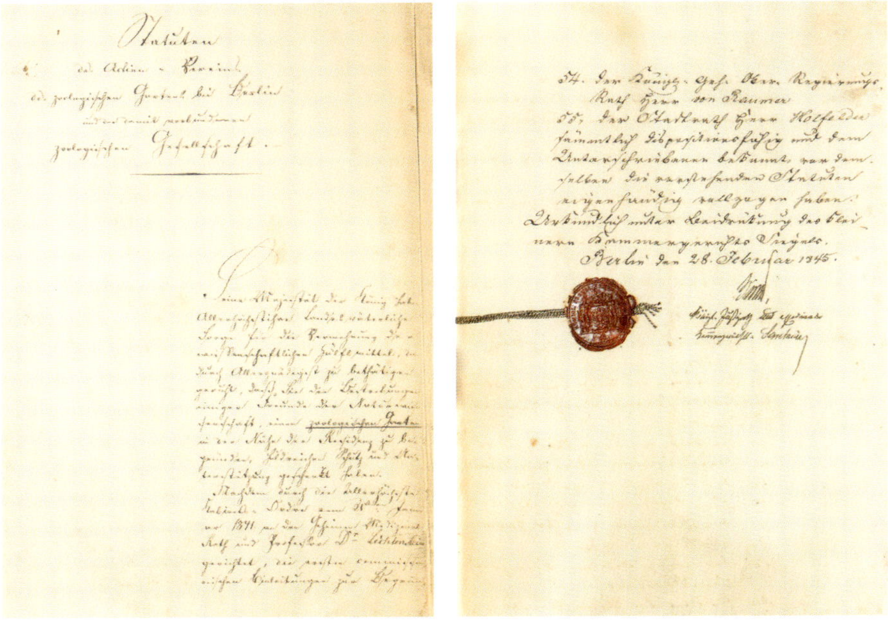

Extract from the statutes of the "Actien-Verein" (shareholders' association) of the Zoological Garden, 1845

21 April 1847
Beginning of the "potato riots"

12 October 1847
Founding of electrical engineering company Siemens & Halske

17 November 1847
First public meeting of Berlin's municipal council

Buildings of the Zoological Garden, circa 1849

Hesitant investors

Founding director Hinrich Lichtenstein had assumed that the Zoo would soon be able to pay for itself. He believed that the best way for it to do so would be to establish a joint-stock company, but this did not yet exist at the time of opening. Until January 1845, the young Zoo was officially run by the royal committee.[17] Then, on 28 February 1845, a group of fifty-five Berlin citizens (fifty-four men and one woman) founded the joint-stock company "Actien-Verein des zoologischen Gartens bei Berlin". The first purchasers of the shares were primarily the city's high-ranking bourgeoisie – bankers, merchants, senior officers and civil servants – but also members of the petite bourgeoisie like pharmacists and tradespeople. Among the shareholders were well-known personalities such as master builder Christian Gottlieb Cantian, historian and heraldist Leopold von Ledebur, and bankers Joseph and Alexander Mendelssohn, the son and grandson of renowned Jewish philosopher Moses Mendelssohn.[18]

Originally, the king had rejected the idea of a joint-stock company since he assumed that such a company would be less geared towards the Zoo's scientific goals and more towards making a profit. However, the statutes of the Actien-Verein,

18 March 1848
Fighting on the barricades as the March Revolution reaches its apex

23 August 1848
First General German Workers' Congress in Berlin

developed during the inaugural meeting, helped Culture Minister Friedrich Eichhorn reassure the monarch.[19] The non-profit structure of the company and its explicitly stated objectives – scientific observation of the animals as well as public and youth education – were quite convincing. Eichhorn also saw the recently enacted laws governing joint-stock companies as a way to keep the Actien-Verein in check. In pursuit of one of its primary goals, the Zoo offered reduced admission prices for schoolchildren and students from vocational schools, grammar schools, cadet schools, orphan schools, schools for the deaf, and charity schools.[20]

The creation of the Actien-Verein was in line with a Europe-wide trend. In the mid-nineteenth century, natural history associations were emerging everywhere, with their founders keen to give citizens a better understanding of divine order and the works of Creation by instilling in them a greater knowledge and appreciation of the natural world. From this perspective, a visit to a zoo could constitute a way to advance one's "moral education". The Zoo's shareholders also supported this concept, and the only "dividends" they received was free admission for themselves and their families. The founding of a joint-stock company was a clear sign that the new zoo in Berlin was at its core a civic project despite the fact that it received state support and was under state control.

However, it also became evident that Berlin was not yet on a par with the wealthy cities of western Europe. There was not yet a large, assertive middle class such as existed in Amsterdam, Antwerp, London and Dublin. Beyond the close-knit group of the Actien-Verein's founding members, the Zoo's shares found very few takers. Of the first 500 share certificates issued, less than 200 were sold before 1868.[21] In time, the shareholders would come to include the Borsig family of industrialists and high-ranking churchman Immanuel Hegel, son of the famous philosopher Georg Friedrich Hegel.[22] But at that time the city and its bourgeoisie were not yet sufficiently developed for the Zoo to truly flourish. It would take two decades for that situation to change.

A site in the sticks

The biggest problem for the new zoo was its location. It is hard to imagine today, but at that time the Zoological Garden "by" Berlin really was quite far from the centre of Berlin and even further away from the neighbourhoods where most Berliners lived. As can be seen from maps dating from around 1850, the Zoo was still very much outside the city even a decade after it was founded, sitting on the southwestern edge of the extensive Tiergarten park.

Wealthier residents who lived in the newer and fancier western neighbourhoods of Wilhelmstadt, Friedrich-Wilhelm-Stadt, Dorotheenstadt and Friedrichstadt had to undertake a long walk from the former city gates – the Brandenburg Gate and the former Potsdam Gate on today's Leipziger Platz – through the Tiergarten to reach the Zoo. People in poorer quarters such as the Spandauer Vorstadt or Stralauer Vorstadt had to cross the entire city, usually on foot. Those who could afford a cab or perhaps even had their own carriage were only a little better off, as the route took them through the still undrained Tiergarten, which would certainly have presented some challenges. One year after the Zoo's opening, the Actien-Verein saw no alternative but to ask the finance minister for "macadamisation" of the road due to complaints about the poor access route.[23] What's more, there were no signs at

1852
Publication of the first definitive German dictionary by the Brothers Grimm in Berlin

15 May 1852
Canal linking the Spree to the Landwehr Canal opens

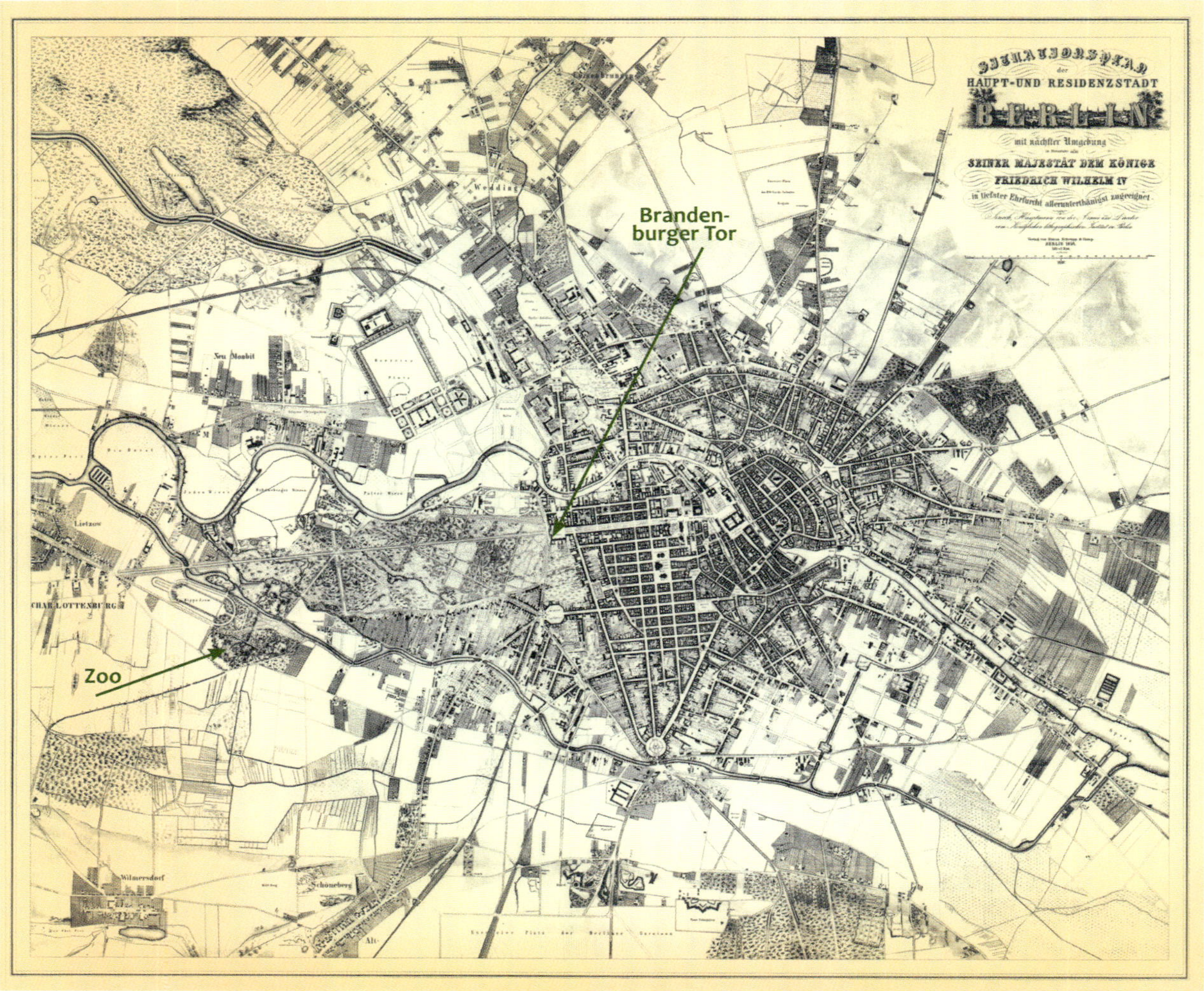

Berlin and its surroundings, 1856

all in the Tiergarten that pointed the way to the Zoo.[24]

A visit to the Zoo by cab was made even more expensive by the fact that passengers would probably have to pay the driver to wait outside the Zoo for the journey back. Records from the Berlin police chief at that time, Eugen von Puttkamer, indicate a set price of twenty *Silbergroschen* for a wait of ninety minutes. This was sufficient time, Puttkamer stated, for a visit to the Zoological Garden "to be accomplished".[25]

Tight budgets

As already mentioned, initially many of the Zoo's animals were housed in makeshift cages and wooden containers. Sturdier dwellings had to be found quickly for the stronger predators. At its opening, the Zoo had a bear pit containing three full-grown "Asian bears" (brown bears) and two juvenile bears "from Lithuania" (probably also brown bears). There were also two "Missuri bears" – i. e. black bears from North

31 May 1851
Unveiling of the equestrian statue of "Frederick the Great" on Unter den Linden

1853
Founding of the Schultheiss brewery

America – kept in a "predator pit" by the western fence.[26]

Captive bears were traditionally kept in pits because of the danger they represent to humans. Usually, these pits were several metres deep, with sheer stone walls. However, such traditional deep pits were not possible at the new zoo due to the high groundwater level. Therefore, in 1847 a new "bear fortress" was built – the Zoo's first stylised structure. This ground-level enclosure was located behind a hill and allowed visitors to observe the animals from the side and above. With its castle-like architecture, this structure reflected a popular aesthetic of the day, and the design was copied several times.

The revolutions that gripped Europe in 1848 didn't only have a negative outcome for the insurgents, they also hit the Zoo hard. After some initial concessions to the revolutionary forces, a reactionary counter-revolution swept across Berlin. The people – educated citizens and workers alike – had neither the time nor the leisure nor the money for a trip to the zoo. In the revolutionary year, the Zoo's revenue of 6,000 thalers was far below the minimum necessary for its continued existence.[27] A long-term positive consequence for the Zoo of the disappointments of 1848, however, was the retreat of many citizens into the private sphere. The failed uprising had shown the people that their political interests could be asserted to only a very limited extent, so instead they focused their energies on economic and cultural self-fulfilment. That attitude ultimately benefitted the Zoo. In the immediate aftermath, however, the king was once again called upon to bail the Zoo out. In 1850 he provided another loan. And yet there was still a severe lack of visitors. During one seven-day period in 1849, the Zoological Garden received no income at all. On another occasion in 1850, it welcomed no visitors for two whole weeks.[28] The king reached repeatedly into his own coffers to save the Zoo – in 1853, 1854, 1857 and 1863.[29] At that time, the joint-stock company was still fully subsidised. New animal arrivals were rare. The first big shipment in 1847 was not followed by another large delivery until 1852. This time, the thirty-five animals were mainly gifts, but their transportation, feed and new cages cost the Zoo several thousand thalers. The transport was organized by the Prussian Consul General in Cairo, Baron von Pentz.[30]

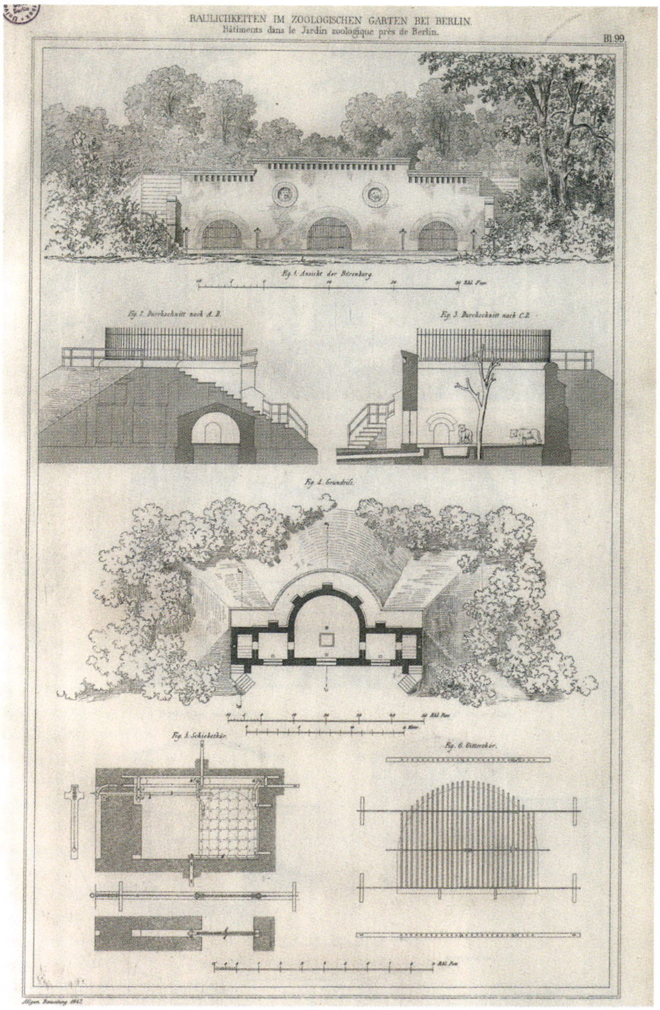

The first "bear fortress" at Zoo Berlin, circa 1847

In 1857, Berlin's first Asian elephant arrived – a male named Boy who had been bought from a menagerie that wanted to get rid of him. After a specially convened board meeting and tough negotiations, the Zoo forked out the not inconsiderable sum of 2,500 thalers for the acquisition.[31] This was seen as an investment in anticipation of an upturn that must surely be on the horizon.

Hinrich Lichtenstein did not live to experience the upturn in person as he passed away in the autumn of that year. His successor was the naturalist Wilhelm Carl Hartwig Peters. Under Peters' leadership, from the mid-1860s visitors could admire new animal attractions, including an alligator. But the Zoo guides published in those decades usually only list single representa-

Zoo Berlin's first elephant, Boy, pictured here circa 1875

2 January 1861
Wilhelm I becomes king of Prussia

1861
The communities of Moabit and Wedding are incorporated into Berlin

23 September 1862
Otto von Bismarck becomes prime minister of Prussia

tives of a species and rarely ever breeding pairs. In the meantime, the Waldschänke ("Woodland Inn") had opened its doors, but even this could not drum up the desired visitor numbers.

It is therefore unsurprising that the younger zoos founded in the rest of Germany quickly surpassed that of the Prussian capital. The zoos in Frankfurt (founded in 1858), Cologne (1860), Dresden (1861) and Hanover (1865) soon enjoyed an equally good or even better reputation than the original role model in Berlin. It was clear that Berlin's zoological garden needed extensive reforms before it could count itself a world-class facility.

The requisite reforms

In 1869, Peters resigned as the Zoo's director. The workload had become too great for someone who was only managing the Zoological Garden on a part-time basis. Before leaving, however, Peters initiated an important administrative reform: his successor should run the Zoo full-time to make the burden more manageable. With the help of committed shareholders, Peters drafted a corresponding amendment to the articles of association, and on 14 May 1869 an extraordinary shareholders' general meeting agreed that from now on there would be one, full-time scientific director who would run the business with the support of a board – which today we would call the supervisory board. Before that, management had been shared among three royal commissioners appointed by the culture minister, each with different areas of responsibility. With the new statute, the influence of the shareholders' meetings also grew, which made owning shares a more attractive prospect.[32]

Following royal approval, the most important reform was introduced: an increase in share capital. Five hundred new shares were issued and the existing shares were exchanged. In 1870 and 1873, fixed-rate bonds were also sold. The Zoo, now operating as the Zoological Garden *of* Berlin, also received permission to clear away some of the trees that had previously been under strict protection.[33] All these measures were intended to increase the attractiveness of the Zoo. Unlike the first issuance of shares back in 1845, the new share sale was a great success. With their share purchases, the citizens of Berlin helped the Zoological Garden raise the funds it so urgently needed for its expansion. By this time, Berlin had also experienced a boom in capital, cultural interest and middle-class philanthropy, meaning that the Zoo could count on broad support from the population and become a genuine attraction. Having long been in the shadow of the younger German zoos, the original zoo in the capital city of Prussia – very soon to become the capital of the German Empire – was now on its way to becoming a popular attraction for visitors from far and wide.

On 1 October 1869, Heinrich Bodinus became the first full-time director of the Zoological Garden. He turned out to be the right man at the right time. Unlike his predecessors, Bodinus already had a considerable amount of zoo experience, as he had been developing the zoo in Cologne since 1859. Zoos were a relatively new concept at that time, so while many of the early directors exhibited plenty of scientific and zoological knowledge, very few had actual practical experience. Bodinus was one of those few, and he soon set about implementing the urgently needed expansion and reorganisation of the Zoological Garden in Berlin.

The creation of the German Empire in 1871 brought about an economic boom, which also

23 May 1863
Founding of the Social Democratic Party of Germany (SPD)

1863
First edition of Brehms *Thierleben* is published

5 September 1866
Inauguration of Berlin's New Synagogue

11 May 1869
Opening of Berlin's first aquarium on Unter den Linden

Heinrich Bodinus (1814–1884), circa 1875

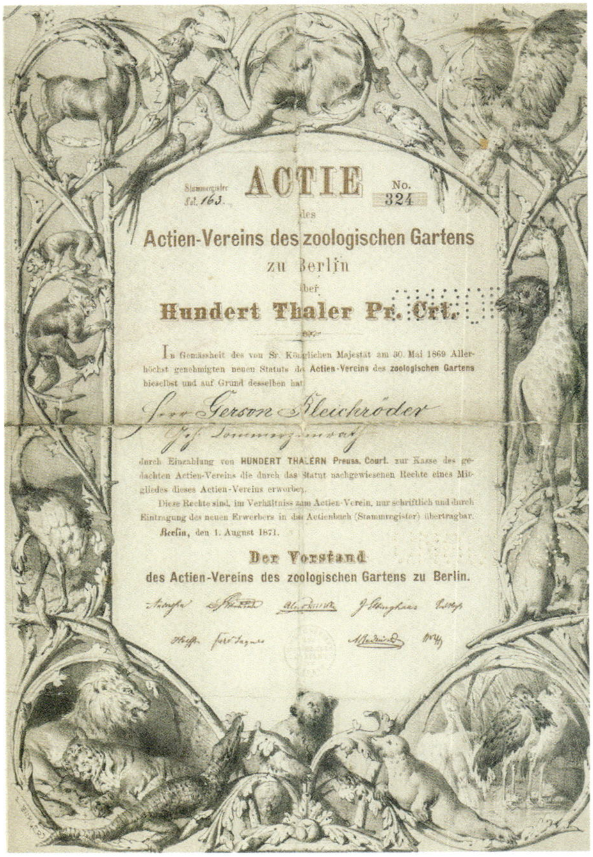

Stammregister Fol. 163.

ACTIE

No. 324

des

Actien-Vereins des zoologischen Gartens

zu Berlin

über

Hundert Thaler Pr. Crt.

In Gemässheit des von Sr. Königlichen Majestät am 30. Mai 1869 Allerhöchst genehmigten neuen Statuts des Actien-Vereins des zoologischen Gartens hieselbst und auf Grund desselben hat

Herrn Gerson Bleichröder

durch Einzahlung von HUNDERT THALERN Preuss. Court. zur Kasse des gedachten Actien-Vereins die durch das Statut nachgewiesenen Rechte eines Mitgliedes dieses Actien-Vereins erworben.

Diese Rechte sind, im Verhältniss zum Actien-Verein, nur schriftlich und durch Eintragung des neuen Erwerbers in das Actienbuch (Stammregister) übertragbar.

Berlin, den 1. August 1871.

Der Vorstand

des Actien-Vereins des zoologischen Gartens zu Berlin.

Share No. 324 of the Zoological Garden's Actien-Verein, owned by Gerson Bleichröder, 1871

greatly benefited the ongoing remodelling of the Zoo. Solvent, generous, committed donors emerged who were very keen to help the Zoological Garden blossom into the scientific and educational institution of international repute that it would eventually become. One early shareholder was the Berlin banker Gerson Bleichröder, who was ennobled in 1872 as one of the few Prussian Jews to receive the honour. The Bleichröder family was part of a group of Jewish people in Berlin who regarded their commitment to the Zoo and other cultural projects as a sign of their integration into German society. The dedication of that family and many other Berlin citizens to the Zoological Garden shows how important social and cultural activities had become for many of the city's inhabitants in the period following the political setbacks of 1848.

Handle with care!

Zoos keep many animals that can be dangerous to humans if encountered in the wild. Anyone who comes into direct contact with these creatures – even those that are accustomed to being around people – must take special care. Zoos have always had rules in place to manage how keepers and visitors interact with the animals. Those rules are continuously updated in line with the latest findings from zoologists and behavioural biologists on species-appropriate animal-keeping, as well as on the basis of keepers' experiences. What seems unthinkable today used to be common practice, and what was once considered normal or even advanced animal-keeping would appal modern-day zoo professionals. Apes, for example, were once given alcohol and cigarettes and forced to sit together at tables and perform tricks, while elephant rides were once a highlight of a visit to the zoo. Modern zoos incorporate the informed opinions of experts and justified public criticism into their work and continuously amend their practices accordingly. Nevertheless, when sufficient care is not taken, incidents can occur between keepers and animals. Usually, any injuries sustained by a keeper or an animal is the result of human error.

There is an old zookeeper saying: you only know whether an enclosure is escape-proof once an animal has escaped. In its early days, the Zoo had a number of break-outs. In 1849, a lion escaped from its provisional enclosure and caused minor injuries to a child. In June 1858, Berlin police filed a complaint with the Zoo, stating that it had not reported the escape of three wolves.[34] Much later, on a night in August 1954, several rhesus macaques escaped. After a prolonged search they were eventually recovered hiding out in the ruins of the nearby Kaiser Wilhelm Memorial Church.[35] In September 1974, female lowland gorilla Cocotte seized the opportunity of an open cage door to escape into the park for ten minutes. During that time, she chased a family before being recaptured by keepers.[36] In 1979, another ape – this time a chimpanzee – found a weak point in his enclosure. His brief excursion ended with a tranquiliser dart shot from

Chimpanzee Missi with cigarettes and a coffee pot, circa 1910

Female chimpanzee Lore on a bicycle, circa 1940

Chimpanzee Suse sitting down to lunch with Mr and Mrs Liebetreu, circa 1940

a blowpipe.[37] In 1983, a brown bear named Petzi escaped her enclosure by cleverly using some overgrown vegetation at the back of the habitat to scale the wall. She scampered across the playground before keepers managed to surround her and the zoo veterinarian stunned her with a tranquiliser dart.[38] Cornelius Schuler, the owner of the Zoo restaurant, sent the director a poem he had written that drew parallels between her escape and the political situation in Berlin at that time:

Berlin's bear, the smartest of all / Secretly jumped over her wall / She didn't want to cause any harm / She just wanted to raise an alarm / Padding around on her big paws / She was demonstrating for a cause / I'm not here to just look pretty / I'm the symbol of this city / I tell you all now, loud and clear / There is no wall for me here![39]

The challenges of keeping exotic animals and the dramatic changes in their housing conditions are demonstrated particularly vividly through Zoo Berlin's history with elephants. In 1857, the male elephant Boy – who, incidentally, was tuskless his whole life – moved into an area in the predators' winter accommodation, which was completely unsuitable and not strong enough to house an elephant. This prompted the building of the Zoo's first elephant house in 1860. However, no one seems to have foreseen an entire herd of elephants living there, as it was only designed for a single animal. When keepers wanted to clean the indoor stable, they could only let Boy out from the inside. In 1867, this situation led to the now almost three-and-a-half-metre-tall elephant actually killing his keeper. Even the magnificent Elephant Pagoda, which opened in 1873, also proved to be unsuitable by later standards. In 1883, an accident occurred involving the Asian elephant bull Rostom. On 28 January, he pinned his keeper Heinrich Krüger against the wall, crushing him to death. It is possible the elephant believed Krüger was trying to take his food.[40]

For a long time, the potential danger represented by the bulls was the biggest obstacle to breeding elephants in captivity. In 1963, another tragic accident involving an elephant occurred at Zoo Berlin. Keeper Günther Lenz was cleaning the outdoor area when the African bush elephant Salim grabbed him with his trunk out of the blue and proceeded to dash him

50 Pf

B.Z.

Die größte Zeitung Berlins

Panik im Berliner Zoo

Braunbär ausgebrochen

Eine Tasse Tee mit dem Chef der Vereinten Nationen

Der höchste Mann der Vereinten Nationen in Berlin: Der fröhliche Schnappschuß, aufgenommen im Garten des Senats-Gästehauses in Grunewald, zeigt UN-Generalsekretär Javier Perez de Cuellar mit Richard von Weizsäcker. Nach einer Tasse Tee, einem ersten Gespräch und einer Ruhepause im Interconti trug sich Perez de Cuellar im Schloß Charlottenburg in das Goldene Buch ein. Beim Essen gab es Krebssuppe, Lendenschnitte und Williamsbirne (Lesen Sie bitte den Bericht auf der Seite 11).

Paul Schockemöhle wieder Deutscher Meister der Springreiter
Seite 70

Wackel-Knie! National-Torwart Toni Schumacher muß am Meniskus operiert werden
Seite 64

SCC kann wieder hoffen! 5:1-Sieg gegen St. Pauli
Seite 67

Kinder schrien, Frauen und Männer flohen

Zoo über eine halbe Stunde lang gesperrt

Das war der aufregendste Sonntag in der langen Geschichte des Berliner Zoos: Ein drei Zentner schwerer Braunbär brach aus seinem Gehege aus, rannte durch Büsche, über Wege und Gartenanlagen. Entsetzen unter den Menschen. 12 000 besuchten an diesem Tag den Zoo.

Der Zoo mußte über eine halbe Stunde lang gesperrt werden. Erst dann gelang es dem Tierarzt, den Bär mit einem Betäubungspfeil außer Gefecht zu setzen.

Der Ausbruch des Bären war möglich, weil das Tier ein Stück der Umgrenzungsmauer herausgebrochen hatte und sich festkrallen konnte.

Lesen Sie bitte den Bericht auf den Seiten 4/5

Front page of the B.Z. *from 13 June 1983 ("Panic at Zoo Berlin: Brown bear breakout")*

against the wall. A second keeper, Robert Jung, succeeded in driving Salim away with a broom, while others dragged the unconscious Lenz to safety. Unfortunately, help came too late and Günther Lenz died of his injuries.[41] At the time, Zoo director Heinz-Georg Klös was on a ship bound for South Africa to collect a pair of white rhinos. His staff notified him of the incident via radio. Klös was deeply shocked. What had made Salim so aggressive? The most pressing question was what to do with the elephant bull now. Should he be isolated or could he stay with the herd? Did he need to be sent away or even put down?

Many people felt very strongly about Salim and the death of the elephant keeper, and Zoo Berlin received letters from all over the country. As is often the case in such situations, the letters were full of "good advice" and reasoning based less on facts and more on emotion. Many implored the Zoo to let Salim live, often projecting human emotions and morals onto the animal. Most of the letters posited that Salim had surely not intentionally killed his keeper, or that Lenz must have provoked him in some way. But there were also letters calling for the removal and even killing of the elephant. Klös and his team certainly had a serious decision to make, as an aggressive male elephant represents a very high risk for all the keepers working with him. Salim was kept in isolation for a while, then reunited with the herd and later kept in his own separate bull enclosure.[42]

Elephant bull Salim, 1958

After the arrival of male elephant Kiba from Houston Zoo in 1997, Zoo Berlin introduced the practice of "protected contact" for the handling of its elephant bulls. This means that any direct contact between animal and keeper takes place through protective barriers and the keepers never enter the enclosure. Since his arrival in Berlin in 2000, the Asian elephant bull Victor has therefore been carefully trained to ensure that keepers and veterinarians are still able to provide him with medical and general care. He has learned, for example, to put his sensitive feet through the protective bars on command so that he can receive foot care, and to present his ear so the veterinarian can take blood samples. The elephants at Zoo Berlin are no longer trained for entertainment purposes.

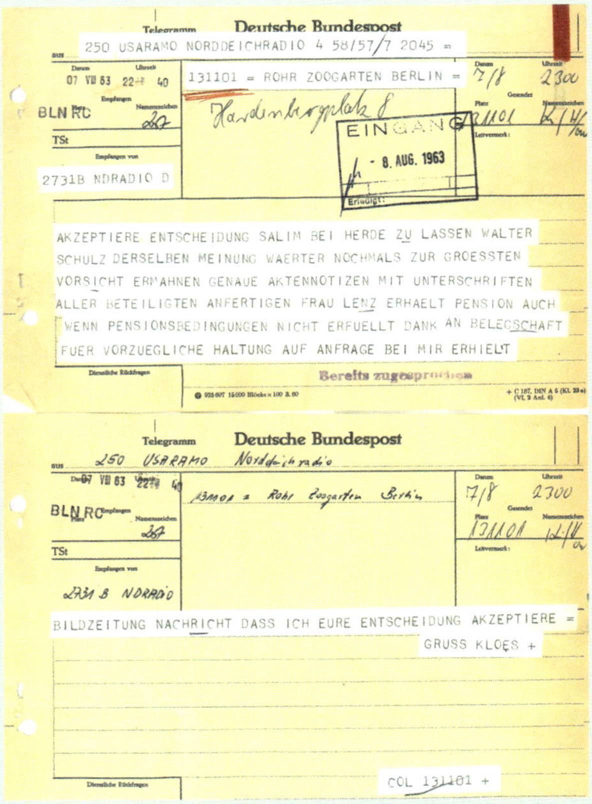

Telegramm Deutsche Bundespost

250 USARAMO NORDDEICHRADIO 4 58/57/7 2045 =

07 VIII 63 22 40

131101 = ROHR ZOOGARTEN BERLIN =

BLN RC

Hardenbergplatz 8

EINGANG - 8. AUG. 1963

2731B NDRADIO D

AKZEPTIERE ENTSCHEIDUNG SALIM BEI HERDE ZU LASSEN WALTER SCHULZ DERSELBEN MEINUNG WAERTER NOCHMALS ZUR GROESSTEN VORSICHT ERMAHNEN GENAUE AKTENNOTIZEN MIT UNTERSCHRIFTEN ALLER BETEILIGTEN ANFERTIGEN FRAU LENZ ERHAELT PENSION AUCH WENN PENSIONSBEDINGUNGEN NICHT ERFUELLT DANK AN BELEGSCHAFT FUER VORZUEGLICHE HALTUNG AUF ANFRAGE BEI MIR ERHIELT

Bereits zugesprochen

Telegramm Deutsche Bundespost

250 USARAMO Norddeichradio

07 VIII 63 22 40

131101 = Rohr Zoogarten Berlin

BLN RC

2731 B NORADIO

BILDZEITUNG NACHRICHT DASS ICH EURE ENTSCHEIDUNG AKZEPTIERE =

GRUSS KLOES +

COL 131101 +

Telegrams from Heinz-Georg Klös to the Zoo regarding Salim, 7 August 1963

The early years of the empire: A city and zoo in bloom, 1870–1900

Clearing the way

Originally founded as a scientific institution, the Zoological Garden of Berlin was now increasingly becoming a place of recreation for the general public. The Zoo's second share issue was a defining moment, as the income it brought enabled the Zoo to acquire more animals and construct additional attractive buildings. Berlin's economic growth and the expansion of the city limits also incorporated the Zoo more closely into the city geographically. From the east, the Berlin suburbs had crept up south of the Tiergarten to the Zoo's gates. And from the west, the town of Charlottenburg was getting ever closer. Citizens of Charlottenburg and Berlin were discovering the Zoological Garden as a source of relaxation and entertainment. Now, as well as a place to observe and admire animals, the Zoo became a location to see and be seen. One of the most important tasks for the Zoo's new director, Heinrich Bodinus, was to redesign the park. The dense trees cast too much shade on the animal houses and enclosures, negatively impacting on the conditions for animal-keeping. The park therefore had to be cleared, as well as drained. This also applied to a hectare of Tiergarten land on the Zoo's northern boundary that had been newly granted to the facility by the Prussian Treasury. Bodinus oversaw alterations to the complex network of paths originally laid out by Peter Joseph Lenné and made considerable changes to the overall structure of the existing park. Those changes included the creation of the Zoo's own "Lake Lucerne", the Neptune Pond, and the Grotto Pond, which all drew a lot of water out of the boggy soil. Those ponds still exist in the southern part of the Zoo grounds today, as a focal point of many beautiful views. The former landscaped park interspersed with the occasional animal enclosure was now transformed into a zoological garden dominated by the enclosures and with a much more moderate quantity of trees and bushes. Natural parkland gave way to well-kept lawns and flowerbeds.[1]

The increase in share capital also aided the realisation of Bodinus's ambitious construction plans. As early as 1867, the board had discussed the creation of a new restaurant. Now that idea was to become a reality. This, plus the orchestra shell built in 1870, significantly heightened the Zoo's appeal as a meeting place for Berlin society.

Rothschild's giraffe Max, 2018

Giraffe

Giraffes (*Giraffa camelopardalis*) are generally divided into nine subspecies, all of which are native to the savannahs of sub-Saharan Africa. Some subspecies are highly endangered, while other populations are more stable. These ruminants can grow up to six metres tall. They spend half the day grazing on bushes and trees with the help of their long necks and prehensile tongues. Depending on subspecies and sex, giraffes need to eat up to 60 kilos of foliage per day.

Female giraffes and their young live in groups of around twenty individuals, while adult males lead solitary lives. A female giraffe's gestation period is 15 months, after which she gives birth to a calf that weighs about 100 kilos. The babies live off their mothers' milk for a year and a half. Around 50 per cent of all young giraffes fall prey to predators – in particular lions and hyenas. Adult giraffes, on the other hand, can defend themselves quite well. A kick from one of their hooves can seriously injure a lion. Giraffe bulls also engage in fierce combat with each other. They use their long necks to thrash their "horned" heads against their opponent's body and legs, or to throw each other off balance. The only time giraffes are completely defenceless is during their phase of deep sleep, which lasts less than half an hour.

These graceful ungulates with their extraordinarily long necks have always fascinated humans. In 1828, the viceroy of Egypt presented the king of France, the king of England and the emperor of Austria with giraffes that had been captured in Darfur. In all three countries, the animals caused a great sensation among the public. The giraffes' appearance even influenced the fashions and women's hairstyles of the time.

Giraffes Mugambi and Max, 2015

Giraffes at the Antelope House, circa 1935

In 1872, the construction of the Antelope House, which also housed giraffes and zebras, represented an important milestone in the Zoo's history. The animals were kept in oval stables with fan-shaped outdoor areas. The Antelope House is the only animal house at Zoo Berlin that was subsequently reconstructed in the style of the nineteenth century original. Today, various sections of the building have been connected inside and out to form larger habitats.

Berliners increasingly perceived the Zoological Garden as a showpiece of their city and flocked there in great numbers. Revenue trebled between 1870 and 1873.[2]

As the new German Empire was being founded, new animal houses were being added in quick succession. In 1870 and 1871, the Large Predator House and the new Bear Fortress were built. The Zoo then gained exotic additions with the Antelope House in 1872 and the temple-like Elephant Pagoda in 1873. The next year, a large concert hall was added, and in the 1880s three open-air concerts were held each week. The Zoological Garden had officially become one of Berlin's top attractions. At last, impressive visitor numbers secured its financial future. Newspapers and other contemporary witnesses would later give Bodinus all the credit for the Zoo's great success, but the decisive measures had actually been initiated by his predecessor, as had the plans for the highly popular restaurant.

E. Arnold, Impressions of the Zoological Garden, *1872*

The large Predator House, circa 1900

There was one slightly bitter pill to swallow alongside all those triumphs: the Zoo was com-

18 January 1871
Proclamation of the German Empire; Wilhelm I becomes emperor

17 June 1871
Part of the Berlin Ringbahn railway line goes into operation

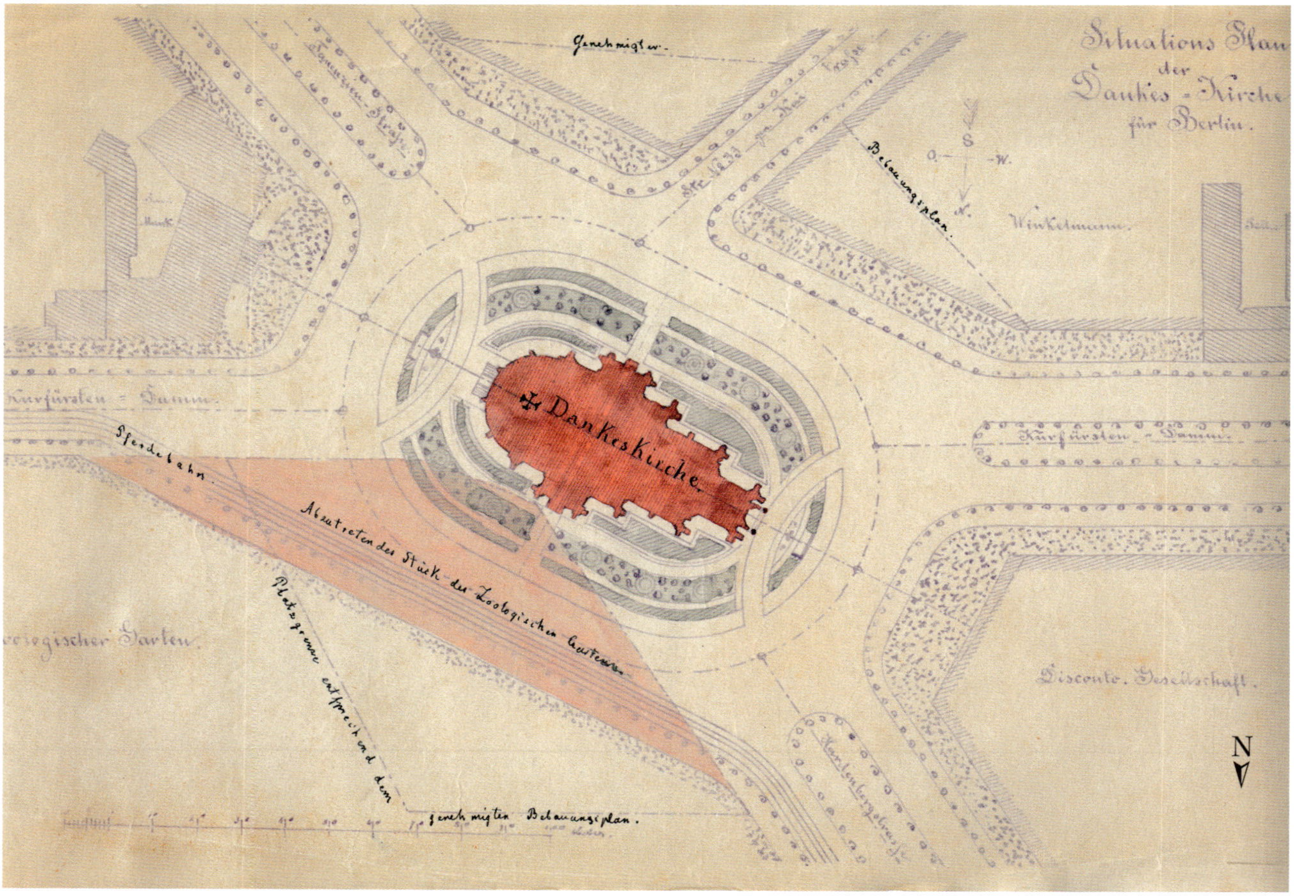

Map showing plans for land changes, 1880

pelled to give up some of its land in the south. As early as 1880, there were plans to build a church as a symbol of thanksgiving for the salvation of Kaiser Wilhelm I, who had survived an assassination attempt two years earlier. The church was to be built on a traffic island in the rapidly developing area around the Kurfürstendamm.[3] That meant the Zoo had to relinquish about 3,700 square metres of land on its southern edge. Initially the Zoo was supposed to give up this land free of charge. But in 1892, after the Zoo had followed the order, it ultimately received compensation from the Imperial Treasury of 20,000 marks.[4]

The Orient in Berlin

At the heart of Zoo Berlin, two minaret-like towers rise above the trees on either side of a glass dome. Then as now, the Antelope House is a popular attraction for visitors and an eye-catching part of many views around the park. Nowadays, people sometimes dismiss the building as an expression of an outdated architectural style – Orientalism – and may even find it kitschy. In the eyes of its contemporaries, however, the building was an architectural sensation. Constructed in under two years by architectural firm Ende & Böckmann, the Antelope House has a floor area

1871
Berlin's population reaches more than 800,000

1873
Imperial Germany adopts the gold standard, replacing the previous currencies in the various German states

of 2,000 square metres. As far as visitors at that time were concerned, the architects had admirably fulfilled their remit of evoking the homeland of the giraffes and antelopes that were housed within. Wild ungulates do not and have never lived in or near mosques, but at a time when only the wealthiest bankers or the highest-ranking diplomats and private individuals could travel to the distant countries where these animals originated, this building and the subsequent ornamental structures at the Zoo probably satisfied the yearning of ordinary visitors for a taste of foreign lands. This was an important function of zoos at that time, as the only other sources of information about exotic parts of the world were written accounts (illustrated, if you were lucky) by researchers and explorers.

It was Bodinus who brought this concept of Orientalist buildings to Berlin. Impressed by the Moorish-style architecture on display in the zoological gardens of Belgium and the Netherlands, Bodinus had already introduced the style at the zoo in Cologne – in the wading bird house, the llama and camel house and, in particular, the giraffe and antelope house. Orientalist architecture also experienced great popularity outside zoos. Kings and princes all over Europe had fairytale palaces built that evoked the Arabian Nights. The

The Antelope House, circa 1920

May 1873
Stock market crash ends
the Wilhelmine economic boom

2 September 1873
Inauguration of the
Victory Column

An antelope, gnu and quagga at Zoo Berlin, 1873

most famous of these is surely the Royal Pavilion in Brighton, built for the Prince Regent, subsequently King George IV of England. In Germany at that time, two large Moorish-style synagogues were inaugurated in Cologne's Glockengasse and Berlin's Oranienburger Strasse, in 1861 and 1866 respectively.

The opening of the Zoo's first Oriental-style building had actually been planned for the year 1870, but its completion was delayed due to a weeks-long workers' strike as well as construction of the large "Lake Lucerne" in front of the building, which was necessary for controlling groundwater levels and therefore had to be completed first. So it was not until May 1872 that the new animal house was opened and the first inhabitants were able to move in.[5]

Originally, the light-filled interior of the Antelope House was also intended to be a sort of tropical greenhouse for exotic plants. However, the air inside the glass-domed hall contained so much ammonia from the animals' urine that the plants could not survive. Later, some of the open façades of the enclosures were glazed, but the hall still proved unsuitable for growing delicate plants, so it was never used as originally intended. Another impractical design feature was the architects' decision not to include a path for the keepers inside the house so as to maximise the space and allow visitors to get closer to the animals. Instead, this walkway was situated between the indoor and outdoor facilities, both of which were elevated. That meant each time the keepers wanted to let the animals out or bring them back in, they had to fashion a bridge from heavy wooden planks. Later, this design fault was corrected.[6]

Thanks to donations and new acquisitions, the Zoo's animal stock grew steadily. It also included zoologically significant individual specimens such as a quagga, which lived at the Zoo in Berlin from 1872 to 1877. This rare equine was probably the penultimate of its kind to ever be shown in a zoo. The only other captive quagga at that time lived at the zoo in Amsterdam, and it died shortly after the one in Berlin. Quaggas had already been hunted to extinction in the wild by humans.

In 1873, architects Ende & Böckmann built a new elephant house. Here, too, the original homeland of the animals was supposed to be reflected in the architecture – this time, India. The skyline in this part of rapidly expanding Berlin was now dominated by the impressive towers of the Elephant Pagoda, the tallest of which was not surpassed until 1895 by the spire of the neighbouring Kaiser Wilhelm Memorial Church. The Elephant Pagoda was intended as a "pachyderm house" that would present various species of this "family" alongside one another. At the building's inauguration, elephants and rhinoceroses – probably white rhinos – moved in. The interior could easily accommodate 500 people. The comfort of the animals had been considered too, as the Elephant Pagoda also had an outdoor area, albeit a small one. This was a first for the elephants in Berlin, one of whom was so frightened by his new-found freedom that at first he could not be persuaded to step outside.[7]

The Elephant Pagoda, circa 1920

As beautiful as the new structure was to look at, there were some serious errors in its design, as knowledge of animal-keeping and the appropriate design for animal houses was not what it is today. First of all, it became apparent that even iron bars with a diameter of eight centimetres are not sturdy enough to contain an elephant, so spikes had to be added to the bars. There was also no separate walkway for the keepers. They had to move between enclosures by squeezing through the gaps between the bars, which were large enough for humans if not for pachyderms. The heavy doors to the outdoor enclosures could only be opened by the elephants themselves pushing against them or by means of a winding mechanism. However, to operate that mechanism, the keepers had to enter the elephants' stalls, which constituted a considerable safety risk.[8] The facility was also only large enough to house one elephant at a time. It was probably not assumed that the Zoo would ever have two African or Asian elephants at the same time. In 1874, Berlin's first hippopotamus also moved into the house.

The Ostrich House designed by architects Kayser & von Grossheim was modelled on an Egyptian temple. It opened in 1901 and housed African ostriches and Australian emus. Both buildings were badly damaged during bombing raids in World War II and subsequently demolished.

Berlin's heightened status as the capital of the German Empire brought many prominent guests to the city, leaving the royal court in need of impressive places to take its high-ranking visitors. Increasingly, the Zoological Garden became the perfect venue. As early as 1872, both Emperor Franz-Joseph I of Austria-Hungary and Tsar Alexander II of Russia were guests of the German Kaiser, Wilhelm I. A visit to the Zoo was on the agenda, and so it came to pass that on 8 September 1872, three of Europe's most powerful rulers enjoyed a carriage ride around the Zoo's parkland before sitting down to a glass of beer in the newly constructed Antelope House. The tsar was so impressed that he gifted the Zoo a pair of European bison. The next head of state to visit the Zoological Garden, Shah Naser al-Din of Persia, came a year later and was guided personally by Director Bodinus. Thus started a tradition that continues to the present day.

The Ostrich House, circa 1930

Interior view of the Ostrich House, circa 1930

The many new ornamental buildings gave the Zoo its characteristic appearance, but the motivation behind their construction was not just the desire to have appealing architecture: the animal houses were also an expression of the philosophies held by the Zoo's directors during that period with regard to the exhibition and collection of animals. The systematic approach that began under Bodinus was continued by his successors, in particular Ludwig Heck.

By the 1880s at the latest, it was established practice for the Zoological Garden to present its animals for study and observation according to their biological order. That allowed visitors to directly compare and contrast a large number of different kinds of animals that shared an evolutionary lineage. The Antelope House opened by Kaiser Wilhelm I in May 1872, which housed antelopes, giraffes and zebras in oval stalls and fan-shaped outdoor enclosures, is one example of this concept. Only the giraffes were provided with larger outdoor areas. At the turn of the century, the Zoo's deer park comprised around 65 enclosures with many different species, and the bird facility contained numerous small cages arranged in three storeys.[9] All this gave the impression of a natural history museum, except the exhibits were alive. Visitors could stroll past the enclosures as if they were display cabinets. The Zoo management wanted to present all the diversity of the world's fauna and allow visitors to gain new insights from comparing the animals with one another. With such an approach, housing conditions were a lesser concern, and so the animals had few opportunities to run around outside. In order to be able to show off as many species as possible, the Zoo also kept many animals as solitary specimens. By today's standards, this would be considered an act of cruelty, especially where the animals concerned are supposed to live in groups and have highly developed social behaviours. The entire Zoological Garden therefore reflected the prevailing understanding of science and education at that time, as well as the predominant attitude that most people had towards animals.

Such tightly packed collections of animals could also represent a danger for everyone in the event of an emergency, but fire safety was not yet a consideration. The Zoo learned a painful lesson in this regard in October 1899, when the Predator House built in 1871 caught fire because of a defective heater. One jaguar was killed in the blaze and two lions sustained minor injuries. Zoo employees and the Zoo's director Ludwig Heck were quickly on site to let the animals into their outdoor cages, while the fire brigade tackled the flames. Thankfully, the new extension had just been completed, meaning the animals could simply move in somewhat earlier than planned. Not as much wood had been used to build this structure, and so it was considered less of a fire hazard.[10]

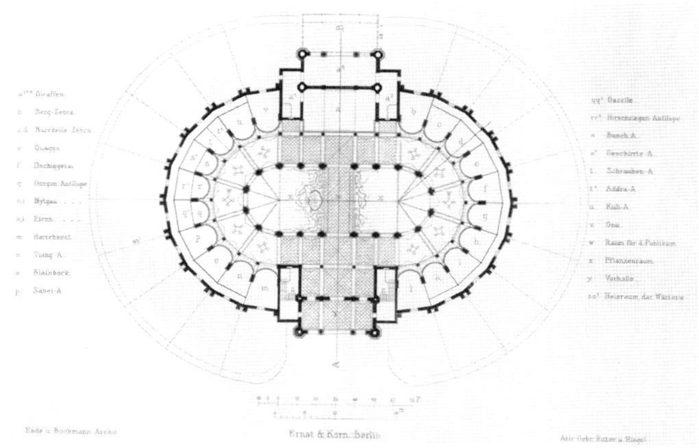

Floor plan of the Antelope House, complete with outdoor enclosures, 1875

Wilhelm Kuhnert, Fire at the Predator House, *1899*

Restaurants and concerts

Over the decades, it became increasingly clear that, as well as the chance to see a wide variety of animals and learn about the natural world, zoo visitors wanted entertainment and culinary

1880
Construction of Anhalter Bahnhof railway station

Concert advertisement poster, 1909

Concert at the large orchestra shell, circa 1915

experiences. In the late 1860s, the management of the Zoological Garden in Berlin began thinking of ways to expand the existing dining options, as visitors did not seem impressed with the modest options available. In 1870, after a little more than a year's construction, a larger and more attractive restaurant opened behind the Neptune Pond. This was another example of how Bodinus was able to take the credit for something that had actually been decided before he became director. But Bodinus did bring one very successful idea with him to Berlin from Cologne: the staging of large concerts. An orchestra shell was built facing the terraces of the new restaurant and was fairly quickly expanded. From May 1870, military music concerts were held here regularly and drew a lot of visitors. It seems that this was precisely the sort of venue Berlin's "high society" had been lacking. Reflecting the zeitgeist of the Wilhelmine era, these concerts featured patriotic march music and anthems such as "Die Wacht am Rhein" ("The Watch along the Rhine").

On the basis of its tremendous success, the large restaurant was expanded still further just two years later. When the grand *Kaisersaal* ("Imperial Hall") opened in 1874, there was no doubting that Berlin's Zoological Garden was truly the place to be. The numerous improvements made the Zoo hugely popular with Berliners from all walks of life. One contemporary observer noted: "A few years ago, most Berliners were hardly even aware that their city had a zoological garden. They did not talk about it, they did not enter it, and those who had visited it had very little praise for the place." A place few people had bothered to tread just a few years before was now a "place of pilgrimage", where everyday citizens would come with their families on Sundays to "experience a few hours in this little paradise. [...] In the past, the highest number of visitors in a day was slightly over 3,000; last summer, the garden had more than 45,000 visitors on a single Sunday." The new, large restaurant had more than 10,000 chairs set up around 5,000 tables – and often even these were not enough. Some days, more than 200 waiters were on duty. In the afternoons in particular, the elegant carriages and cabs of the few wealthy citizens would cause a traffic jam at the entrance facing the Tiergarten. "Thanks to Bodinus, the Zoological Garden has become a favourite destination for Berliners, a draw for people

The Kaisersaal *in the large restaurant complex, circa 1910*

from elsewhere."[11] From 1872, the Zoo was considered the "jewel of the royal city".[12]

Even so, the Zoo did not rest on its laurels. For example, when Asian elephant Boy died in 1879, the management had his skeleton prepared by the conservator of the Berlin Zoological Museum. From 1883, the skeleton was on display in the Elephant Pagoda. The white colour of the huge bones tempted some visitors to leave their "names and other personal notes" on it, much to the dismay of the Zoo management.[13] This kind of "graffiti" is still a problem at the Zoo to this day.

Skeleton of Boy the elephant inside the Elephant Pagoda, 1883

The city comes to the Zoo

Part of the reason for the Zoological Garden's increased popularity and significance was its greatly improved accessibility. On 22 June 1865, the first horse-drawn tram in Germany went into operation in Berlin. These vehicles drawn on rails were considerably more comfortable than those that traversed the roads on ordinary wheels. The tram line ran from Kupfergraben at

1881
First public bicycle races take place in Berlin

1 April 1881
Start of telephone traffic in Berlin, with around 50 users

Lichtensteinallee entrance in the Tiergarten, circa 1910

the tip of today's Museum Island along Charlottenburger Chaussee (today's Strasse des 17. Juni) and through the Tiergarten, connecting Berlin with the independent city of Charlottenburg. This significantly shortened the route through the Tiergarten to the Zoo. Later, a separate line was established to take people from the Grosser Stern square at the heart of the Tiergarten to the Zoo entrance. A terminus with a stable building was created here in 1874 on the Zoo's grounds, but outside the fence. Here, horses could rest before being re-harnessed for the return journey.[14] The site had been made available for this purpose by the Prussian Treasury. From July 1875, trams stopped directly in front of the entrance every 20 to 24 minutes.[15] The Zoo could now be reached by visitors within an acceptable length of time for the price of just 2 ½ *Silbergroschen*. In 1880, another line was added that ran from the "Knie" (today's Ernst-Reuter-Platz) via Kurfürstenstrasse and terminated very close to today's Elephant Gate (*Elefantentor*).[16]

But not everyone could afford to travel on the new mode of transport. For most working people who lived in the new tenements that were now sprouting up everywhere, a visit to the Zoological Garden remained a rare luxury. But the growing class of civil servants and white-collar workers, who were working diligently to improve their standing and education in an effort to emulate the bourgeoisie, were increasingly drawn to the Zoo.

The original gate located in the Tiergarten behind the Landwehr Canal gradually became regarded as the "back entrance" to the Zoo. The new transport routes connecting Berlin, Charlottenburg and Wilmersdorf were on the Zoo's western and southern edges, which led to a need for new gates. In the late 1870s one such gate opened on Budapester Strasse, which was still called Kurfürstenstrasse at that time. In addition to the new connections, an entire railway station had sprung up on the western edge. However, it seems that the importance of the new transport route was not yet properly understood by the board of the shareholders' association. In January 1878 and again in April 1879, the management of Berlin's city railway contacted the Zoo's management to inform them that it would be advisable to install a new gate facing the soon-to-be-opened railway station. When the Zoo's board had still not responded shortly before the station was due to open in November 1881, the railway management got in touch again.[17] They explained that the station had been designed as big as it was to accommodate the large numbers of visitors they expected to be thronging to the Zoo, "particularly on certain summer days".[18]

The Zoo management then quickly responded by opening a provisional station-side entrance with two ticket booths. The new city railway line was a further stroke of good fortune for the Zoo. From February 1882, it connected Schlesischer Bahnhof in the east of the city (now Ostbahnhof) with the railway station in Charlottenburg to the west. The twelve-kilometre, nine-station line thus linked up the Zoo to the city of Berlin. Although a

May 1881
Berlin's first electric tram goes into operation

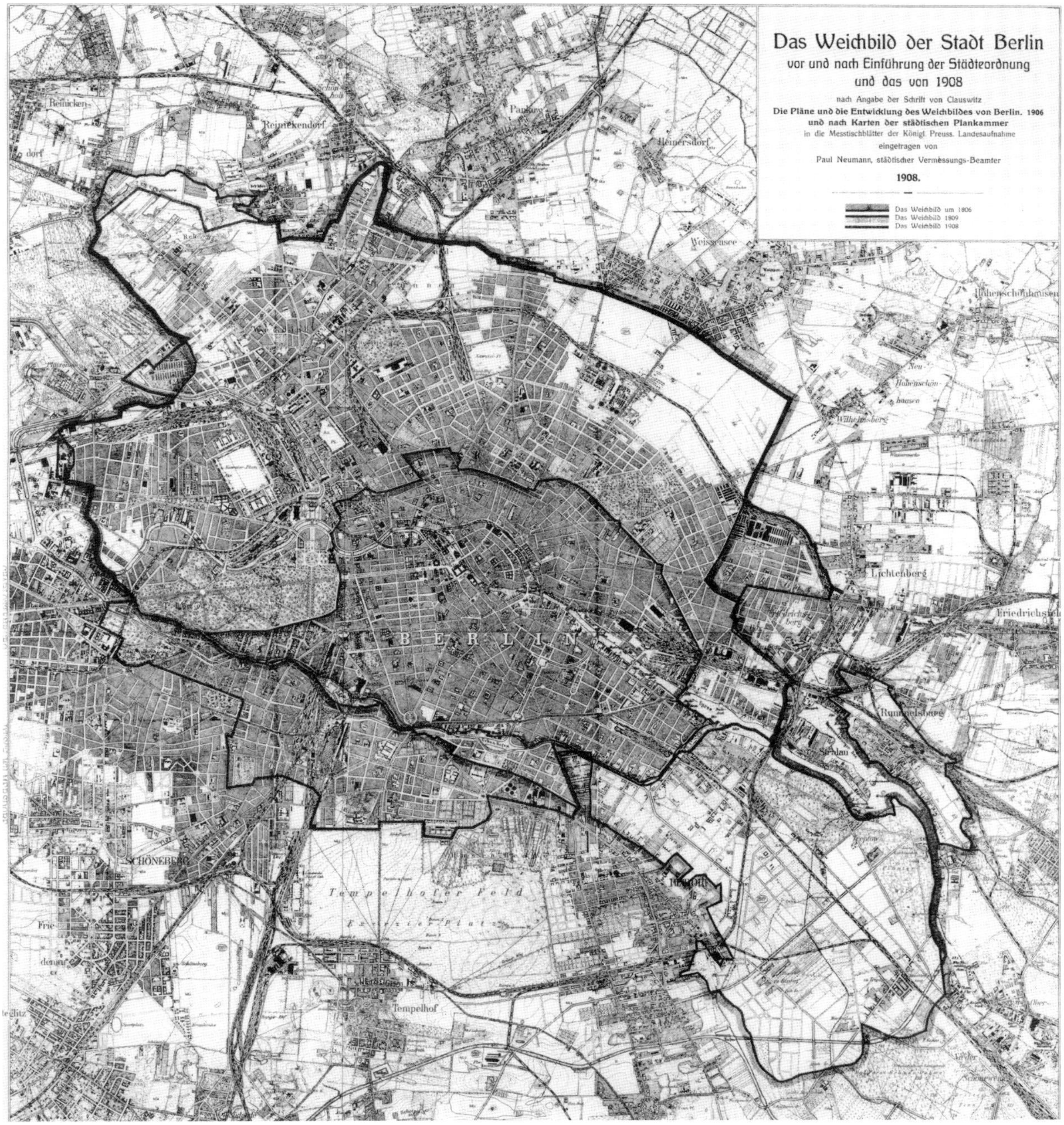

Map of the city of Berlin, 1908

railway ticket remained something of a luxury for many Berliners for some time to come, the Zoo's clientele became more diverse and visitor numbers kept growing.

Soon, the crowds arriving via the railway station grew so large that Zoo director Ludwig Heck decided to have a more prestigious entrance built in place of the provisional one. Architects Zaar

& Vahl designed square ticket booths, topped by large stone lions with a row of carved animal heads beneath them. With its new Lion Gate (*Löwentor*), the Zoo now gave visitors travelling by rail a proper welcome with a more distinguished portal. As expected, the project went over budget, probably partly because the Prussian Finance Ministry wanted a "handsome" structure next to the new railway station, and so the gate turned out more monumental than originally foreseen.[19] The new entrance became even more important when the Zoologischer Garten underground railway station opened in 1909. Although the later Elephant Gate on the south-eastern edge would ultimately become more famous and is now considered the main entrance by most Berliners and tourists, more visitors still enter the Zoo through the Lion Gate facing the "Bahnhof Zoo" station.

The growth of Wilmersdorf and Charlottenburg also influenced the Zoo's development. Soon, large, imposing buildings were being erected south of the Zoo on Kurfürstenstrasse and the Kurfürstendamm. Then, the tram line on Kurfürstenstrasse was electrified.[20] That soon made the horse-drawn tram terminus on the Zoo's north-eastern edge superfluous – and, anyway, the Zoo now needed the land that housed the depot and stables for its ever-expanding collection of animals. Shortly after the turn of the century, following years of negotiations, it regained the area thanks to the support of the Prussian Finance Ministry.[21] German industrialisation – which came rather late by international standards but

Günther Klebes, Steam-powered tram in front of the Elephant Pagoda, *circa 1886*

was now rapidly gaining momentum – coupled with Berlin's increasing prominence as the capital of a great European power meant that the population was also increasing, along with the amount of available capital. As a result, the towns of Berlin, Charlottenburg and Wilmersdorf were advancing ever closer to the Zoo. The population of Charlottenburg tripled between 1875 and 1890 to more than 75,000, while Wilmersdorf doubled in size with more than 5,000 inhabitants.[22]

Over time, the area south of the Tiergarten attracted many affluent citizens. Many Zoo share-

The Lion Gate with a sign pointing to the underground station, circa 1909

Subscription cards for the Zoological Garden, circa 1900

holders lived in Charlottenburg, Tiergarten and Wilmersdorf – which would all later become boroughs of Greater Berlin. The Zoo had started selling season passes shortly after it first opened, and these "subscription cards" became de rigueur for wealthy Charlottenburgers and Berliners. In the evenings, the social elite would promenade along the Zoo's avenues. Lists of subscribers from the years around the turn of the century read like a who's who of the educated bourgeoisie – the ones who could afford these cards and were keen to have them.

However, if the Zoological Garden wanted to live up to its prestigious reputation and call itself a genuine place of education for the people, it had to enable those on lower incomes to visit the Zoo as well. Director Heinrich Bodinus therefore made the decision to halve the admission price on the first Sunday of each month. On these "low-cost Sundays", much larger sections of the population could visit the Zoo for just 25 pfennigs. Contemporary media reports and photographs show that the Zoo was always very well attended on the "low-cost Sundays". Of course, the clientele

15 June 1888
Kaiser Friedrich III dies; Wilhelm II becomes the new emperor

Emil Koch, 25-Pfennig Sunday at Zoo Berlin, *1882*

was rather different to that of weekdays, when subscribers and visitors from outside the city would saunter along the main promenade. Now larger groups, including many children, crowded in front of the cages. Instead of dining opulently in the restaurants, the lower classes would bring along packages of sandwiches, which they would wash down with beer.

Heck the collector takes the helm

On 23 November 1884, Heinrich Bodinus died unexpectedly of heart failure, cut down in the middle of a busy, active life. He left behind a zoological institution that had long outgrown its humble beginnings. The Zoological Garden of Berlin had an important standing in the city and in German society. This was certainly due in part to the increased size and importance of Berlin itself, but also to the hard work of the Zoo's director Bodinus, who had doubled the number of animals and tripled the variety of species on display.

His successor Maximilian Schmidt only got to experience three years as Zoo director before passing away in office. Schmidt's main achievement was overseeing the construction of the new Hippo House. The Moorish-style brick building had been planned for some time, as the Elephant Pagoda only had space for a single hippopotamus. The Hippo House was the Zoo's first building dedicated to a single animal type. The Elephant Pagoda, Monkey House, Predator House and Antelope House had always been intended for multiple species.

With Schmidt's death in 1888, the position of director at Berlin's zoo was vacant once more.

1891
Otto Lilienthal conducts his first successful gliding attempts

1895
Inauguration of the Kaiser Wilhelm Memorial Church

Applications came in from all over the German Empire. Two final candidates were selected: Ludwig Wunderlich, director of the zoo in Frankfurt, and Ludwig Heck, director of the Cologne zoo. The latter came highly recommended by his former professor in Leipzig, the parasitologist Rudolf Leuckert.[23] Wunderlich did not have references to equal this. And so for the second time in its history, the Berlin Zoological Garden found itself in the hands of a man from Cologne.

Heck was very keen to expand the Zoo's animal collection. He wanted to exhibit as many species as possible and thus place Berlin's facility on a par with the most important zoological institutions in Europe. By 1899, his ambitious goal was realised: with 1,215 species of mammals and birds, the Zoological Garden of Berlin was the world's premier zoo in terms of species diversity.

However, Heck got off to a bumpy start. After just one year, he became embroiled in an argument with the new emperor, Wilhelm II. The point of contention was the planned killing of the elephant bull Rostom in 1889/90. Rostom had demonstrated very aggressive behaviour; in 1883 he had even fatally wounded his keeper Heinrich Krüger. Heck's predecessors had already fortified the elephant's enclosure and forbidden anyone from riding him. We now know that Rostom's aggressive tendencies and some of his other behaviours would have been typical responses to severe shortcomings in his keeping. In any case, the idea of putting Rostom to sleep was a delicate matter, as he and fellow elephant Omar had come to Berlin in April 1881 as a gift from the Prince of Wales. The prince, later to become King Edward VII of England, was Wilhelm's uncle, the son of Queen Victoria. Victoria, who had herself gifted a female elephant to the Zoo in 1862, also intervened in the matter. The Prince of Wales strictly forbade the killing of Rostom, paying no heed to arguments that Rostom had exhibited violent behaviour and had even previously killed a keeper in London in 1879, prior to his journey to Berlin. Although Wilhelm II was quite generous when it came to financing the Zoo, he was never one of its biggest fans – and he certainly didn't want any trouble in his extended family. Unlike his wife Auguste Viktoria, Wilhelm II never officially visited the Zoo.[24] Rostom finally died from an illness in 1892.

Following the completion of the many new buildings in the 1870s, there had been a break in the Zoo's construction activities. Given Heck's dedicated commitment to expanding the facility into a "systematic zoo" by collecting more and more species, in the 1890s new animal accommodation was required. Equipped with a secure financial basis and supported by patrons and donors, Heck launched an extensive construction programme.

In 1884, the park received electric lighting and the paths were paved – the old sand paths had become unbearably dusty given the more than 80,000 visitors some Sundays.[25] Now, the home of the feathered residents was due an expansion. The previous aviary, built in 1872, was no longer large enough. Following the extension and rede-

The old Hippo House, circa 1890

1897
Inauguration of the Kaiser Wilhelm national monument on Schlossfreiheit square

1899
First art exhibition by the Berlin Secession group

The new Bird House, circa 1910

Interior view of the new Bird House, circa 1930

sign of the existing structure, a magnificent new Bird House opened in June 1895, just in time for the summer season. Like other Zoo buildings of the period, it was designed in the Moorish style, with the chimney from the heating system concealed inside a stylised minaret. The building housed more than 400 types of birds, including 100 parrot species. A keepers' corridor divided the interior into two separate public areas. The larger and particularly interesting species such as hornbills could be admired on the front side in separate, glass-fronted units designed to resemble the birds' natural habitats. There were also individual heatable glass vivariums for the more sensitive species located in the middle of each visitor area. Countless smaller cages ran along the interior walls in three "storeys", which could be accessed and attended to from the keepers' corridor behind them.

The collection resembled nothing so much as supermarket shelves or a library stocked with living exotic birds. The cages were organised in line with the standard classification system of the time – a clear demonstration of the director's systematic scientific approach. Heck wanted ornithologists, bird lovers, and regular visitors to gain an impression of the evolutionary history and diversity of the animal kingdom. With so many birds in such an enclosed space, the sound must have been quite overpowering, not to mention the smell. A special ventilation system was developed to mitigate that problem.[26]

In 1896/97, a Japanese-style Stork House was erected in the immediate vicinity of the Bird House. Just one and a half years later, the magnificent Elephant Gate, also in the Japanese style, opened at the Zoo's Budapester Strasse entrance. It was supplemented by an administration building and operational facilities. Thanks to the redesigned pathways, early twentieth century visitors

1900
Physicist Max Planck gives his first lecture on quantum physics

5 August 1901
Death of the popular empress Kaiserin Victoria

The Elephant Gate, circa 1910

who walked through this portal were greeted by the beautiful Dreisternpromenade, which led directly to the Chinese Music Pavilion. Original designs had a fountain positioned in the centre of the gate. However, this would have created an obstruction, and so the architects opted instead for the impressive – albeit not very Japanese – elephant sculptures. Following a faithful post-war reconstruction of the originals, these elephants still adorn the gate to this day, providing a hugely popular photo backdrop.[27]

New structures next to the director's residence included the Deer House, a Pheasantry, and an area for mountain goats and sheep. Those facilities also consisted of multiple small, separate enclosures that literally showcased Heck's extensive range of animals and systematic categorisation of species. These, too, were meant to allow for comparative study of the various animals and their characteristics by Zoo visitors. To properly achieve this goal, the Zoo needed to house as many different species and subspecies as possible. Ludwig Heck was not the only one to adopt this approach. In zoos all over Europe, animals of numerous different species were presented alongside one another like museum exhibits. The animals – in many cases creatures with pronounced social behaviours and needs – were frequently displayed as isolated specimens, while the zoos competed against one another to boast the greatest collection of species and the most extensive classification of animals.[28]

1901
Nobel Prizes are awarded to Berlin-based physiologist Emil von Behring and physicist Wilhelm Conrad Röntgen

1904
The Tietz department store opens on Alexanderplatz

1904
Opening of the Bode Museum on Museum Island

Rudolf Virchow examines the members of a Sámi group, 1879

Abraham Ulrikab and family, circa 1880

Human zoos: "Ethnological exhibitions" at Zoo Berlin

From the beginning of the Age of Discovery, groups of people from forcibly subjugated territories were repeatedly put on display in the royal courts of European colonial powers.[29] In the second half of the nineteenth century, private commercial operators began recruiting, transporting and exhibiting people and cultures that were exotic to Europeans. These impresarios claimed that the shows were intended to widen visitors' knowledge about cultures in other parts of the world, but in reality they were mainly geared towards sensationalism. The European audiences generally had preconceived notions of these "savages" in their heads – images that had been propagated by press articles and novels. In order for the organisers of the shows to have financial success, the "performers" had to meet the audience's expectations. Sometimes, the individuals on show were portrayed as "natives", "primitives" or "innocents". Other times, they were "wild warriors", "dangerous Amazons", "noble savages" or "exotic beauties".

The first such exhibition at the Zoo in Berlin was held in 1878 and comprised a small group of Inuit people – colloquially known as "Eskimos" – from Greenland. Initial concerns about presenting humans alongside animals were dispelled by the fact that Rudolf Virchow, the most important German physician and pathologist of the time, and the Berlin Society for Anthropology that he had co-founded, placed high scientific value on these and subsequent shows. The anthropologists took measurements of the people on display and attempted to use physical characteristics to determine a supposedly "high" or "low" level of development in the respective "races". The fact that these examinations were often humiliating to the individuals involved was of little concern to the scientists.

And yet the most convincing argument in favour of the Inuit exhibition as far as the Zoo's director Heinrich Bodinus was concerned was almost certainly that it had been a great success at the Jardin d'Acclimatation amusement park in Paris. The organiser, famous animal trader Carl Hagenbeck of Hamburg, guaranteed that it would bring in at least the average income for the planned exhibition month of February and that any additional profit would be split fifty-fifty.[30] That was a very tempting prospect, particularly given that the Zoo's cash registers were feeling the strain of the recent construction investments.

After the "Eskimo" show, which was indeed a roaring success, the Zoo then displayed a group of Nubian people from northeast Africa. That show also included elephants, rhinos and other African animals. Once again, it was a huge financial success and firmly established "ethnological exhibitions" as a feature at the Zoological Garden in Berlin.[31] The following year, a group of Sámi people from northern Scandinavia – known at that time as "Laplanders" – were displayed at the Zoo alongside some reindeer. The Zoo also held a second Nubian exhibition.

Most of the groups that came to Berlin travelled all over Europe, appearing at various zoos and show venues. The impresarios who were responsible for taking the bookings and supervising the groups were often also involved in animal trading. Participating zoos saw themselves merely as hosts and did not feel any responsibility for the well-being of the people who were brought to them for the shows.

In 1880, Carl Hagenbeck brought another group from the far north to Berlin, this time two Inuit families from the Labrador region that totalled eight individuals. One of the two adult men in the group, Abraham Ulrikab, kept a diary during his trip. He wrote that he and the others in his group found the big cities frightening and oppressive. On the ensuing tour through several cities, all members of the group fell ill with smallpox and died due to lack of vaccination.[32]

Despite this tragedy, Hagenbeck organised another show the following year featuring a group of indigenous people from the Tierra del Fuego archipelago at the southern tip of South America. It is safe to assume that these people did not come to Europe voluntarily. The exhibition was shown at various locations from October to December 1881, including at the Zoological Garden in Berlin. During that time, the show's participants were unable to communicate with anyone as no one else spoke their language.[33] It would therefore

Erste Daheim-Beilage zu No. 6. 1882.

Aus der Zeit — für die Zeit.

Die Feuerländer.

Mann, Profil.

Kind.

Grönländer und Lappländer, Patagonier und Nubier sind bereits bei uns zu sehen gewesen und nachgerade werden wir allmählich eine Völkergalerie bei uns bewundern können, der zoologische Garten wird teilweise zu einem anthropologischen, der Mensch studiert den Menschen und erkennt, wie verschiedenartig das Ebenbild Gottes gestaltet ist. Aber bei aller Verschiedenheit leuchtet doch wieder die Einheit hindurch. Was an Farbe und Haar, Schädelbildung und Sprache alle die Rassen und Stämme auch unterscheiden mag — hoch darüber steht das allgemein Menschliche. Sie alle sehen und schlafen, essen und trinken, denken und fühlen wie wir, die wir uns als die am höchsten stehenden dünken, wiewohl grad- und stufenweise verschieden. Auch die Feuerländer oder Pescherähs, welche jetzt im zoologischen Garten in Berlin zu sehen sind, und die man für eine ungemein niedrigstehende Menschenrasse ausgab, sind gar nicht so übel und gehören keineswegs zu den häßlichsten Leuten. Es sind die südlichsten Menschen unserer Erde, welche das sogenannte „Feuerland" bewohnen, einen vielfach zerklüfteten, felsigen Archipel, der durch die Magellanstraße von Südamerika geschieden wird. Ihre körperlichen Eigenschaften stellen sie zu der großen amerikanischen Rasse als deren südlichste Ausläufer sie anzusehen sind; von ihrer Sprache ist noch sehr wenig bekannt. Aber nicht alle Feuerländer sind unter sich gleich, auch scheinen verschiedene Sprachen und Dialekte unter ihnen zu herrschen. Es gibt große, kräftige Leute und kleine kümmerliche unter ihnen.

Die Feuerländer im zoologischen Garten zu Berlin.

Jene, die der bekannte Naturforscher Darwin an Ort und Stelle kennen lernte, erschienen ihm als die merkwürdigsten und interessantesten Leute, die er je gesehen. „Ich hätte nicht geglaubt, schreibt er, daß der Unterschied zwischen dem wilden und zivilisierten Menschen so groß sei. Er ist größer als zwischen einem wilden und einem gezähmten Tiere, da der Mensch größere Fähigkeit besitzt, sich zu vervollkommnen." Es waren die östlichen Feuerländer, die er kennen lernte, sechs Fuß hoch, den Patagoniern verwandte Leute. Ihr einziges Kleid bestand aus einem Mantel aus Guanakohaut mit der Wolle nach außen; diesen trugen sie über die Schulter geworfen, wobei ihr Körper nackt blieb. Die Haut ist schmutzig kupferrot, das Haar grob, schwarz, verworren und das Gesicht mit roten, weißen und schwarzen Streifen bemalt. „Die Gesellschaft glich ganz den Teufeln, wie sie im Freischütz auf die Bühne kommen." Ihre Sprache verglich Darwin mit einem harten Glucksen und Räuspern. Sie sind ausgezeichnete Mimiker; so oft die Engländer husteten oder gähnten oder sonst eine linkische Bewegung machten, ahmten sie dieselbe augenblicklich nach. Einige Matrosen fingen an zu schielen und Gesichter zu schneiden, aber ein junger Feuerländer schnitt noch weit gräßlichere Grimassen. Mit vollkommener Richtigkeit wiederholten sie jedes Wort in einem vorgesprochenen Satze und behielten denselben auch einige Zeit lang. Über den Gesang der Engländer waren sie im höchsten Grade erstaunt, nicht minder über das Tanzen. Obgleich sie selten Europäer gesehen hatten, kannten und fürchteten sie doch die Feuergewehre und nichts konnte sie vermögen, eines derselben anzugreifen. Fische, Schaltiere, Guanakos bilden ihre Nahrung.

Die Hütten dieser Feuerländer gleichen in Größe und Gestalt einem Heuschober. Sie bestehen aus einigen wenigen abgebrochenen Zweigen, die in den Boden gesteckt und unvollkommen auf einer Seite mit Grasbüscheln und Binsen ausgestopft werden. Es ist somit nur eine Art Windschirm. Das Ganze kann nur einige Stunden Zeit zum Erbauen kosten, und wird auch nur wenige Tage benutzt. Dabei erinnere man sich, daß das Klima ein sehr schlechtes dort ist und einen großen Teil des Jahres über Schnee fällt, wiewohl die Breite nicht höher ist als das südliche Schweden sie besitzt.

Weit jammervoller ist das Bild, welches von den westlichen Feuerländern uns entworfen wird. Darwin sagt, es seien die jämmerlichsten und elendesten Geschöpfe, die er je gesehen. Sie besaßen nur ein Seehundsfell oder gar nur eine Otterhaut, kaum größer als ein Taschentuch, zur Bedeckung. Andere, die in ihren Kanoes zum Schiffe kamen,

Mann, von vorne.

Frau.

Exhibition of people from Tierra del Fuego at the Zoological Garden, 1882

have been impossible for them to voice any dissent. There were certainly people at that time who suspected this was a case of human trafficking.[34] But the group's "agent" Carl Hagenbeck proudly touted them to Zoo director Bodinus as "real-life cannibals" – adding that the designation "man-eaters" had positively influenced ticket sales at the zoo in Paris. The show was also a great success in Berlin.[35] To avoid further trouble, in December Hagenbeck insisted that the group be accommodated in "covered quarters" when the weather became "too harsh".[36] During the day, however, they had to be outdoors. Despite these supposed precautions, some members of the group started displaying symptoms of lung disease when they were in Zurich, one of the tour locations after Berlin. Hagenbeck cancelled the tour, but some members of the group still lost their lives. The survivors were handed over to a British mission in Chile.[37] Whether the "performers" were ever fully aware of what was

Map of the Zoological Garden from around 1927, with the special exhibition (Sonderschau) area visible in the northwest corner

happening to them remains unclear. In any case, that was of little concern to the participating zoos, which transferred all responsibility for the proper recruitment, welfare and payment of these people to the organisers with whom they shared the revenue. For the Zoo in Berlin, the show was, at least, a financial success, and it brought in a large amount of additional revenue that year.[38]

From that point on, the "ethnological exhibitions" held at the Zoo became increasingly professional: participants usually underwent medical examinations, were specially recruited for the trip, and performed rehearsed choreographies or acted out scenes. However, this does not mean that the Zoo paid any more attention to their well-being and health, especially as it was still just one stop on a larger tour. Sometimes, the supervisors or impresarios would physically punish the "performers" if they did not behave as instructed. The Zoo's management turned a blind eye to this; they just wanted the show to be a success and for order to be maintained.

The professionalisation of the shows' organisation and performances was accompanied by the professionalisation of the Zoo as a host. Towards the end of the century, a dedicated exhibition area in the southwest corner of the Zoo made it easier to control visitor flow and to settle accounts with the show organisers. Later, an arena with terraced seating was built on an extension area in the northwest corner. The first show held there was animal trader John Hagenbeck's "India Show", which featured nine elephants and a very large group of people including acrobats, snake charmers, sword swallowers and fire eaters.[39]

One particularly interesting exhibition from a historical perspective was the Samoan show, which visited several German cities in 1900/1901 and appeared at the Zoo in Berlin for several weeks in the summer of 1900. A show featuring Samoans had already toured Europe from 1895 to 1897. That first show consisted mainly of scantily clad "exotic beauties" and was intended to attract a predominantly male audience. The Berlin venue for that show had been the Passage-

Panoptikum. It is likely that the organisers were now planning a similar show for 1900, as they had recruited predominantly young women and part of the advertising focused on the alluring feminine beauty to be found in that exotic paradise. But by the turn of the century, Samoa had become a colony of the German Empire and the organisers were banned from solely focusing on the erotic representation of women. Instead, the members of the group were now hailed as "new compatriots".[40] Nevertheless, the posters still time in Berlin. The show's organiser had to ensure that the Samoans were at the exhibition site for the duration of all showtimes stipulated by the Zoo's director and that the troupe were kept "under the strictest supervision" by "a white-skinned gentleman familiar with the customs of the Samoans".[41] The foreign visitors were forced to comply with the instructions of the supposedly superior whites. The Zoo's management emphasised that the Samoans' visit was conducive to German colonial aspirations, as it gave them

Advertising material for the Samoan show, 1900

featured the image of an innocent "Eve" with a snake covering her chest. The fact there are no snakes in Samoa was clearly irrelevant.

The group was even presented to Kaiser Wilhelm II. It can be assumed that the Samoan "performers" hoped the trip would foster important contacts with Europeans. However, as usual, the contract between the Zoo and the organiser of this "ethnological exhibition" contained very strict rules for the troupe during their an impression of Germany's power.[42] Not long after, the exhibiting of colonial subjects was banned in Germany. The Samoan show was therefore the only such exhibition at the Zoo that had a direct link with colonial politics.

In 1927, the Zoo tried its hand at organising its own show for the first time. As Germany no longer had any colonies, the Zoo reached out to Italy's Ministry of the Colonies in an attempt to acquire a group of Bedouins.

The ministry's initial response was: "The Arabs are no worse people than ourselves and cannot be exhibited in zoological gardens."[43] But clearly the concerns of the Fascist authorities only went so far, because shortly thereafter they agreed to send a group from Tripoli, provided the show and its advertising highlighted the influence of Italian culture and civilisation on the colonies where Bedouins lived. The Zoo's management assured the ministry that this would be done. After long negotiations, including with the German foreign ministry, the necessary travel permits were issued.

However, audience interest was low and the show flopped.[44] The Zoo's board of directors began to wonder whether the time for such exhibitions might be over. Perhaps, after decades of these shows, the demand was no longer there. After all, most Berliners were now aware of the people and customs on display through pictures and the emergence of cinema, which gave them more convenient and comfortable access to foreign cultures.[45] When Zoo Frankfurt and the Society for Anthropology, Ethnology and Prehistory offered the Berlin Zoological Garden a group of "Chulupi Indians", which they believed would be "extremely appealing for European tastes" and included a large number of "well-developed girls" from Bolivia, the Zoo declined.[46]

In the late stages of the "ethnological exhibition" era, organisers were under pressure to present something truly extraordinary. In 1931, the Zoo put on a show featuring Sara-Kaba people – an ethnic group native to areas of Central Africa south of Lake Chad. At that time, Sara-Kaba women wore wooden discs in their lower lips as ornamentation. The press advertised these women as "Lippennegerinnen" ("lip negresses"). The show also featured some "pygmies", who may not actually have even been members of the same ethnic group; their role was purely to increase the "extraordinary" nature of the exhibition.[47]

Although the event still generated a net profit, it was not hugely successful and was not even mentioned in

The "mosque" from the Tripoli show, 1927

Arrival of the Sara-Kaba people in Berlin, 1931

the Zoo's 1931 annual report. In 1932, the special exhibition area was finally converted into a petting zoo. The concept of the "human zoo" thus lasted longer than it should have and underwent its own development during that time. The "performers" did not always come to Europe voluntarily and, especially in the early days, were sold false hopes of money and recognition. Many were swindled and some even died of unfamiliar European diseases. The harsh treatment frequently practised by the impresarios varied from brutality to neglect. Only later did the groups regularly feature trained performance artists.

From the point of view of the ethnologists who legitimised the shows, the human research objects enriched the science of the day while providing the public with a rare opportunity to view people from foreign cultures. However, the show organisers prohibited personal interactions, which were also discouraged by the authorities. Where such encounters did occur, these were very brief and/or strongly influenced by hierarchical attitudes. The emotional impact of the shows was also significant. Ultimately, they served to reinforce the racist and hierarchical views prevalent among the scientists and impresarios involved of "more evolved" and "less evolved" ethnic groups – views that were passed on to large swathes of the population. Despite some brief moments of contact between the audience and the participants from overseas, these shows were never a meeting of equals, but rather a form of entertainment that reinforced existing stereotypes and gave the viewers only what they wanted to see.

None of the approximately twenty-five "ethnological exhibitions" held at the Zoo really provided any deeper scientific understanding; they were purely about boosting visitor numbers and ticket sales. Ultimately, these dehumanising exhibitions were only possible within the colonial power structures and attitudes of their time.

Global fame and global crises, 1900–1932

On top of the world

Around the turn of the century, the Zoological Garden in Berlin was continually upgrading its parkland as well as constructing new enclosures and animal houses. In 1893/94, irrigation and drainage were improved further and new ponds were dug. Additional buildings appeared almost every year: in 1901 it was the new Ostrich House in the design of an Egyptian temple, in 1903/04 the Nordic Deer Log Cabin, in 1905 the Bison House. Next came the Pig House, and from 1909 to 1910 an ensemble for equine animals was built – the African Animal House and Persian Tower – to designs by Carl Zaar and Rudolf Vahl. Following the tradition of large Orientalist structures at the Zoo, the Persian Tower was reminiscent of a minaret on a Persian mosque, adorned with a domed roof and faience frieze. With its battlements, loggia and wooden balcony, the African Animal House recalled the Islamic architecture of eastern Africa.

The Eagle Rock was also built at this time. The new outdoor aviaries for birds of prey were set around what appeared to be a large rock rather than a traditional animal house. Inside were heated winter quarters for the birds as well as the necessary maintenance and food preparation facilities. A large number of different raptors inhabited the Eagle Rock. The large birds were given a "flight path" that allowed them to remain airborne over a distance of more than 50 metres. The Eagle Rock was considered another of the Zoo's architectural gems. It was badly damaged in World War II but subsequently rebuilt. It was then redesigned in the late 1980s, and in 2018 it was turned into the new Eagle Canyon, with its large walk-through free-flight aviaries.

Other pioneering structures included the waterfowl aviaries, which were built at the turn of the century between the main utilities and administration area and a house for wading birds. A stream ran through the aviaries, and at one end visitors could observe a wide variety of aquatic birds diving for food. The windows of the visitor grotto offered a direct view of the pool occupied by herons, penguins and other waterbirds. The fact that these birds were not found in the same part of the world was wilfully ignored.

In summer 1905, an event of considerable zoological significance occurred in Berlin: the birth of a giraffe. Attempts to breed the large ungulates

A smooth-fronted caiman, 2018

A spectacled caiman, 2019

Crocodile

The crocodilian order (*Crocodilia*) is usually divided into three families comprising more than twenty species in total: alligators/caimans, gharials, and true crocodiles. All these reptiles inhabit rivers and lakes in the tropics and subtropics. Only the saltwater crocodile, commonly found on the coasts of Australia and various islands of Southeast Asia, can also live in the sea. Alongside birds, crocodilians are the only modern-day survivors of the once huge group known as archosaurs, which included dinosaurs and pterosaurs. Crocodilians are ambush predators and perfectly adapted to their aquatic habitat. The poikilothermic reptiles lurk motionless just below the surface of the water, observing their surroundings with their eyes conveniently positioned on top of their heads. Their nostrils, too, are set high on their snout, allowing them to breathe while the rest of their body is almost completely submerged. When prey comes within reach, the huge reptiles strike by propelling themselves forward at high speed with the help of their laterally compressed tail. Crocodilians often kill their catch by holding it underwater and drowning it. One exception is the gharial, which is only found in India and Nepal and mainly feeds on fish.

All crocodilians are equipped with long rows of pointed teeth and incredibly strong jaw muscles. While the adults have few natural predators, their eggs are highly vulnerable to theft by nest predators. Up to 90 per cent of a crocodilian's egg clutch and hatchlings fall victim to carnivorous mammals and birds.

The primaeval power of these reptiles has always fascinated people around the globe. In some cultures, they have even embodied deities. They also probably served as one inspiration for dragons in the mythology of the people alongside whom they live, doubtlessly in part due to the fact that certain species represent a significant danger to humans. The crocodile is ranked the third most deadly animal to humans in Africa, after the lion and the hippopotamus. However, this is primarily due to the fact that the expansion of human settlements and shrinking of natural habitats is bringing humans into ever-closer contact with the reptiles. Today, most crocodilian species are endangered.

At Zoo Berlin, crocodiles have always been the star attraction of the most important structure built on its grounds after the turn of the twentieth century: the Aquarium.

Asha the gharial, 2019

in captivity were rarely successful at that time. The Zoological Garden had begun its efforts in 1900. The female calf was born in the Antelope House and had to be raised by hand by her keepers. Around that time, the Zoo's director Ludwig Heck also succeeded in exhibiting exotic species rarely or never before shown in Germany, such as the greater glider, the Tasmanian wolf and the aye-aye.[1] While the masses were drawn to the baby giraffe, these other animals were of greater interest for more zoologically knowledgeable visitors. Both achievements demonstrate the Zoo's success in the early years of the twentieth century.

Persian Tower of the architectural ensemble for equine animals, circa 1920

The Eagle Rock, circa 1920

Additional animals came to the Zoo from the German Empire's newly acquired colonies. Lieutenant Hans Dominik of the colonial troops (*Schutztruppe*) in Cameroon, for example, brought a young elephant to Berlin. In the years prior to that, Dominik had been using brutal means to suppress uprisings by African peoples. His gift to the Zoo caused a great deal of excitement because this particular elephant had been born in the area of present-day Cameroon and was therefore considered a "citizen of the German Empire". Similarly, a feline that came to the Zoo from southeast Africa in 1900 was described as a "German East African wildcat".[2]

But there was trouble ahead. In 1907, the leading animal trader in Europe – possibly the world – Carl Hagenbeck opened his own animal park, throwing down the gauntlet to existing "systematic zoos". Located in Stellingen near Hamburg, Hagenbeck's *Thierpark* ("animal park") presented "panoramas" of the animal kingdom instead of strictly organised individual specimens. Various species that inhabited the same climate zone lived within a single landscape with dividing trenches, regardless of their taxonomy. It didn't matter to Hagenbeck whether or not the animals would be in such close proximity in the wild. African ungulates grazed alongside their greatest foes, the lions, in peaceful harmony, separated only by ditches that were not visible to onlookers.

Hagenbeck's concept thrilled the public, but it greatly concerned zoologists at established zoos. Most of them, first and foremost Ludwig Heck, considered Hagenbeck's approach unscientific and "theatrical". Even more concerning were Hagenbeck's threats of expansion. In 1911 he was hatching plans to establish another zoo on the Jungfernheide, a large area of wooded heathland northwest of Berlin. He even succeeded in win-

Russian-style log cabin at the European bison enclosure, circa 1930

Card celebrating forty years' employment of head keeper Pechler, 1910

ning Kaiser Wilhelm over to the project. During a visit to Stellingen, the German emperor expressed enthusiasm about a model of the planned new animal park, which was to be situated next to Plötzensee lake, where Afrikanische Strasse and Kameruner Strasse ("African Street" and "Cameroon Street") now meet.[3] At 40 hectares, the animal park was to be twice the size of the existing Zoo. The Zoological Garden's shareholders' association submitted a letter of protest to the culture ministry. Such a zoo was unscientific, it claimed: it was based on inaccurate sensationalism and would only appeal to "a low sort of audience". In addition, the shareholders stated, two zoological gardens could not exist in Berlin side by side.[4] Yet Hagenbeck enjoyed the unqualified support of the Kaiser, and Berlin's city councillors were also inclined to support the project. The incorporation of large areas to the northwest of Berlin into a single "Greater Berlin" was being debated at that time, and the presence of an animal park would certainly boost the area's development.

However, in the longer term, scientific criticism of Hagenbeck's concept by other zoo directors and an increasingly sceptical attitude among Berlin politicians discouraged the Hamburg entrepreneur, and the project ultimately failed.

The Zoological Garden in Berlin had no further competition to fear, at least not for the foreseeable future.[5] In August 1914, with the outbreak of war, it was still one of the world's leading zoos, boasting an inventory of around 3,500 animals of 1,500 species.

Feeding and feasting

Despite all the progress the Zoo had been making, it still needed a stable source of income besides ticket sales. The management and supervisory boards were well aware of the positive effect the leasing of eateries had on the company's budget. Around 1900, the Zoo management began developing plans to use a strip of land alongside Hardenbergplatz for the construction of an exhibition hall, which it could then lease for a profit. Although this would result in the loss of some enclosures and leave the Elephant Pagoda overshadowed by a taller building, the financial security it would bring was just too tempting. In 1902, once the Prussian Finance Ministry's concerns that the new structure had nothing to do with the Zoo's main purpose were laid to rest, a series of halls were erected. A separate company took over management of those buildings. The intended purpose of the neo-Gothic halls quickly proved unprofitable, and they were soon converted into cinemas and entertainment venues. Nevertheless, the new halls were a huge asset for the Zoo, which did not have to participate in any of the administration or organisational work but gained regular passive income from their lease and payment by subsequent users of the space. This additional profit enabled the Zoo to finally secure its financial independence.[6]

The Zoological Garden of Berlin had increasingly become a centre of cultural life in the city. Dressed in their finery, the upper echelons of Berlin society would saunter along the Dreisternpromenade that had been laid out in 1900, past the Chinese Pavilion, and around the piazzas in front of the Music Pavilion. Although many zoologists probably frowned upon this development, it was an inescapable fact that the recreational function of European zoos was gaining significance – not least at the one in the German Empire's capital Berlin.

The fact that the Berlin Zoological Garden was establishing itself as a place of leisure as well as education was apparent in the presence of more appealing eateries. Even just a few weeks after

Exhibition halls at the Zoo, circa 1910

The Dreisternpromenade, circa 1910

1908
Women are officially permitted to study at the University of Berlin

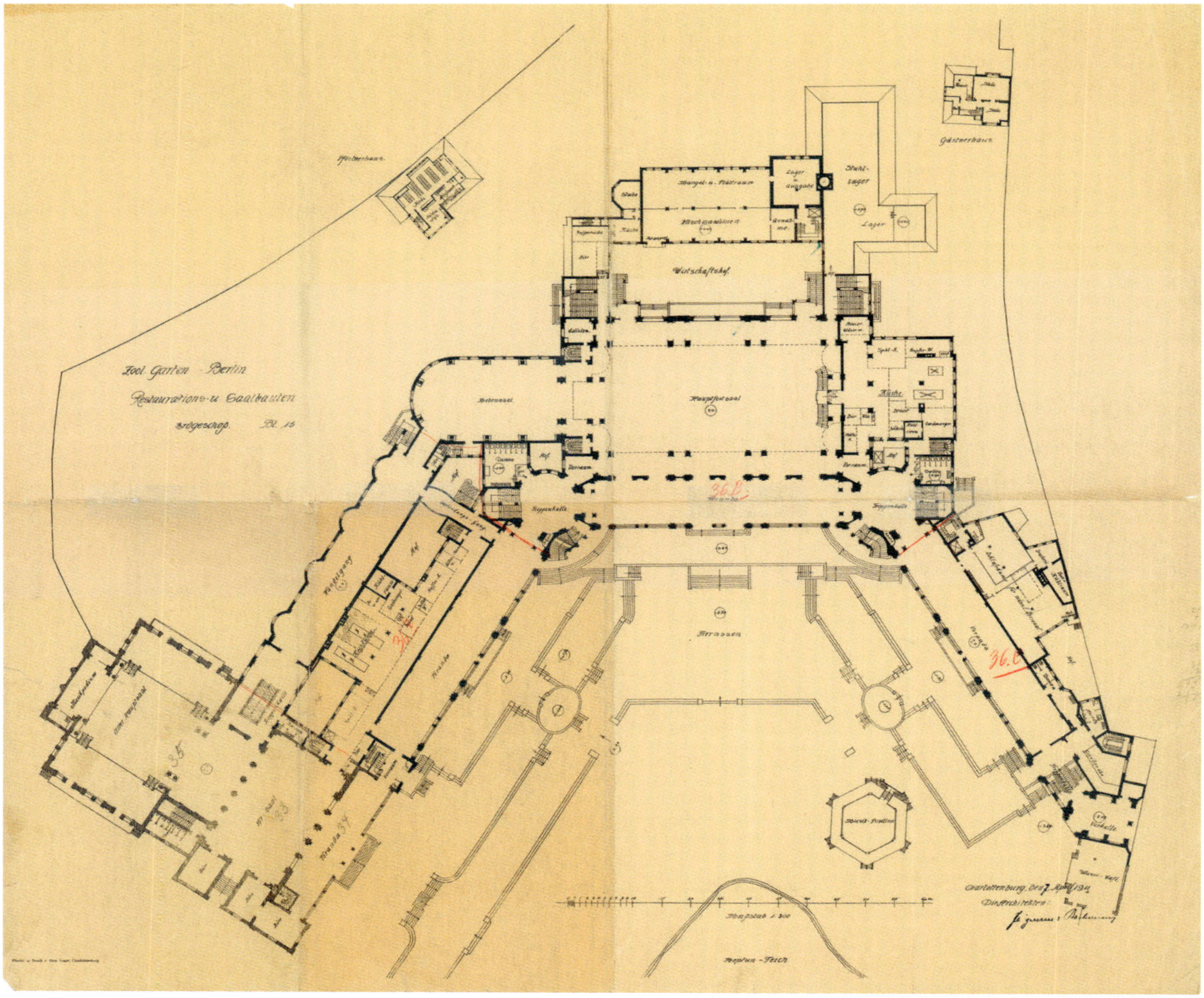

Floor plan of the large restaurant complex, circa 1910

the Zoo opened, visitors had frequently enquired about the options for refreshment. The first eatery, named the Waldschänke or "Woodland Inn", and the subsequent larger restaurant had been very well received, and since the Zoo collected a share of the income of those establishments, catering became part of its business model. The provision of food and drink was such a success that the Zoo management began considering how it could expand those options further. A bidding process started in 1907/08, a provider was selected, and over the following years an ensemble was created that ended up being the largest public catering establishment in Germany to date. Incorporating existing structures from 1874 such as the *Kaisersaal* ("Imperial Hall"), the new ensemble had three wings housing several dining halls, galleries, lounges and terraces. It had six large kitchens, its own bakehouse and a laundry, allowing it to serve 10,000 guests in the halls and the same number in the outdoor area. A separate entrance, the *Adlerportal* ("Eagle Portal"),

1909
First flight operations on the parade ground at Tempelhof

1909
Construction of the Pergamon Museum begins

At the "Thirsty Flamingo", 1903 (top); poster advertising the Opera Ball held in the large restaurant complex (bottom)

allowed exclusive access from the bustling Kurfürstendamm. Illuminated columns visually extended the building into the garden, providing a festive atmosphere in summer and during ball season in the winter.[7] The building cost almost three million marks and opened in 1911.

Soon, these grand rooms were hosting some of the most important balls in Berlin, including the Opera Ball and the Press Ball. Visitors who could not afford the prices in the large, fancy restaurant complex could still enjoy good plain fare

G. Mützel, M'pungu the gorilla, who lived in the aquarium on Unter den Linden

Karl Heyn, inside Brehm's aquarium on Unter den Linden, 1869

and beer at the older, cheaper inn, which operated under the name "Thirsty Flamingo" at that time.

The Zoo gets its own aquarium

Berlin had been home to an aquarium since 1869. Located on Unter den Linden, not far from Friedrichstrasse, it was founded by renowned zoologist Alfred Brehm and was well known throughout Germany, in particular for its remarkable interior design that featured numerous grottos. Brehm had come to the field of zoology relatively late in life, but he quickly gained fame due to the tremendous popularity of his travel reports and magazine articles; every child knew the author of the *Brehms Thierleben* ("Brehm's Animal Life") series of books. In the late 1860s, Brehm was asked to found an aquarium for a specially created joint-stock company. He chose not to follow the taxonomical approach that was common in zoos and natural history museums at that time, and instead divided his aquarium into separate habitats: desert, forest, freshwater and saltwater. Large light shafts illuminated the central aviary, the "Atlantic show" and the "geological grotto". Visitors moved through dark cave-like rooms seemingly carved from natural stone, gazing into illuminated pools and enclosures. From the beginning, the aquarium exhibited birds and reptiles as well as fish and marine animals. Even the first live gorilla exhibited in Europe, M'pungu, lived at the aquarium from 1876. However, he shared the fate of many apes captured at that time and died after just one and a half years in Berlin.[8]

In 1874, Brehm left Berlin, and the pharmacist and zoologist Otto Hermes became director of the aquarium. Hermes led the aquarium successfully for many years but it ran into financial difficulties after the turn of the century. Urgently required restoration work called for larger investments, which the company could not (or did not want to) pay for. Also, the neighbourhood around the aquarium had grown increasingly popular and expensive, which meant tempting offers had been made to purchase the land where the aquarium stood. The director of the Berlin Zoological Garden, Ludwig Heck, had become aware of the aquarium's financial difficulties in 1907 and had begun negotiations with Hermes to take over the facility. However, in early 1909 those negotiations failed due to differences of opinion between the two men.[9]

Finally, on 30 August 1910, the Zoo's board of directors asked the German emperor and king of Prussia, Wilhelm II, for permission to increase the share capital so that it could build its own aquarium. The board members had decided the Zoo should have an aquarium on its grounds because, they asserted, the aquarium on Unter den Linden was no longer viable and negotiations between the Zoo and the Aquarium's owners had not borne fruit. The board's plan was to increase the share capital of 2.8 million marks by 500,000 marks by issuing new shares worth 1,000 marks for the price of 1,250 marks. For an additional 100 marks, existing shareholders could acquire the right to free admission. Permission to proceed was quickly granted, and the share issue took place that same year.[10]

Oskar Heinroth, Director Heck's scientific assistant, was commissioned with planning the

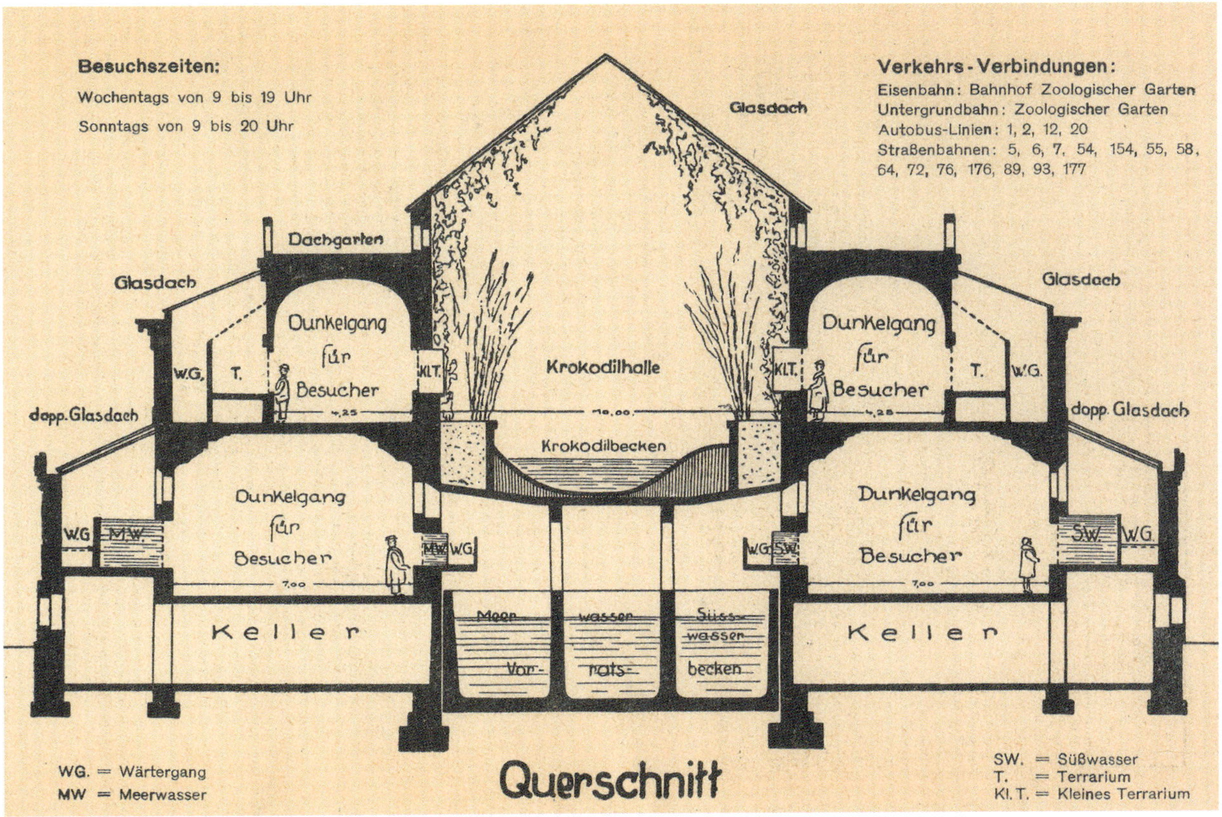

Cross-section of the Aquarium building, circa 1925

1 April 1912
Initial steps taken towards establishing Greater Berlin

Construction of the Aquarium at Zoo Berlin, circa 1912

On the basis of his findings, he conceived of a three-storey building that would house freshwater and saltwater tanks on the ground floor, terrariums for reptiles and amphibians on the first floor, and an insectarium on the top floor. He thus aimed to create much more than just an aquarium: the plan was for it to function almost as an extension to the Zoo, enriching the already extensive collection of fauna from around the world. The new aquarium would be the first place in Europe to devote an entire department to insects and arachnids, rather than just a small niche. Also, to present the evolutionary history

The Aquarium at Zoo Berlin, circa 1913

new aquarium. Otto Hermes died in September 1910. Shortly after his death, the old aquarium building was sold and its animals were transferred to zoos in Frankfurt and Leipzig. The first task Heinroth set himself was to visit some of the largest aquariums in Germany and abroad.

of the animals on display, the aquarium building was to be embellished with reliefs and mosaics of fossils and extinct predecessors of modern-day reptiles and fish. There would also be an impressive statue of an iguanodon (a dinosaur species) outside the entrance from the Zoo grounds.

Aquarium printed guide, circa 1925

The Crocodile Hall following its opening, circa 1913

The supervisory board had ambitious plans to build an aquarium even larger than the existing ones in Amsterdam and Antwerp. The scale of the project was one of the reasons that the costs for Berlin's new aquarium ended up far exceeding the sum provided by the share capital increase.[11] In 1911, the buildings on the site of the Zoo's former power station and coal yard were cleared so that construction could begin. One remaining water tower was integrated into the new structure; it can still be seen today at the southwest corner of the Aquarium on Budapester Strasse. The upper floors were each set back slightly from the one beneath so that skylights could be installed for the tanks and terrariums positioned against the outer walls on the two lower floors. At the heart of the building was the Crocodile Hall. This was to resemble a natural jungle landscape with a river running through it. With its glass roof, it was the world's first open walk-through indoor enclosure. Thanks to the innovative architecture, visitors could already observe the reptiles in the water from the two stairwells – through glass panes set into the walls. Once they entered the Crocodile Hall on the first floor, visitors crossed a creek on a bamboo bridge, surrounded by dense vegetation on all sides. From there, they could admire crocodiles, alligators and gharials floating

1914
More than 7,000 cars are now registered in Berlin

in the water or lying on the banks beneath them. The walk-through enclosure gave visitors an immediate experience of the reptiles' natural environment – the heat, the humidity, the smell of the water, the earth and the animals themselves.

The new Aquarium Berlin opened on 18 August 1913, after less than three years' construction work. It represented the most important expansion of the Zoological Garden to date.[12] The building was designed by architects Zaar & Vahl, who had already completed numerous commissions for the Zoo. Oskar Heinroth, who had overseen the general planning, took over the position of director and made his new facility one of the most renowned aquariums in the world. Heinroth's ornithological research also attracted international attention to the Zoo and its Aquarium.

Barges on the Landwehr Canal unloading seawater for the Aquarium, 1913

Initially, the saltwater tanks were filled with real seawater from the North Sea. Barges transported 350 tonnes of the saltwater along the River Elbe to the Havel and the Spree, and then to the Landwehr Canal and on to the Zoo. Once it arrived, the water was pumped into the Aquarium through a one-kilometre pipe, with the help of the Berlin fire brigade. That process alone took almost twenty-four hours. The transport costs involved were so high that Heinroth soon developed a method for producing artificial seawater.[13]

The newly appointed Aquarium director, Oskar Heinroth, was born near Mainz on 1 March 1871. As a schoolboy, he developed an early interest in zoology through observing native birds.

Aquarium director Oskar Heinroth in his workshop in the Aquarium's attic space, circa 1917

Oskar Heinroth with Ural owls, circa 1929

Young Oskar would catch animals and keep them in his bedroom, in terrariums and tanks he had built himself. He even raised chicks and made some initial breeding attempts. As a grammar school pupil in Dresden, he spent so much time observing duckponds that he earned the nickname "Entenheinrich" (roughly equivalent to "duck boy"). In accordance with his father's wishes, Heinroth initially studied human medicine. Only once he had complemented this first degree with a doctorate in 1895 did he begin to follow his true passion and move on to study zoology. In Berlin, he made the acquaintance of Ludwig Heck, who immediately took him under his wing. Heck allowed Heinroth to observe the Zoo's animals and dissect those that had died. After completing his zoology studies, Heinroth was offered a voluntary position as an assistant at the Zoo. He then spent some time lecturing and took a long trip to Africa, but upon his return he was employed as Heck's assistant, and represented the director in his absence.[14]

In academic circles, Oskar Heinroth is generally regarded as one of the founders of ethology, the scientific study of animal behaviour. This has been confirmed by Konrad Lorenz, the man that most people regard as the father of ethology. Heinroth laid the foundations for this field of study with a short piece of writing on the "ethology and psychology" of waterfowl, which he presented at the International Ornithological Congress of 1910. In the piece, he explained the systematic observations he had made of his beloved ducks. Heinroth was a meticulous man who carefully observed a large number of individual creatures over very long periods of time. His most important work is probably the multi-volume standard work *Die Vögel Mitteleuropas* ("The Birds of Central Europe"), written with his first wife Magdalena.[15]

The Zoo goes to war

The Zoo and Aquarium were not able to enjoy their glory days for long. The assassination of the heir to the Austrian throne, Archduke Franz Ferdinand, in Sarajevo by Bosnian revolutionaries on 28 June 1914 was the spark that ignited the powder keg: on 28 July the Austro-Hungarian Empire declared war on Serbia, triggering a devastating conflict across Europe and the world. On 18 August, the Zoo supervisory board debated closing the Aquarium, which was still running at a loss, in light of "the current circumstances".[16] During the first six months of the war, both Zoo and Aquarium Berlin experienced great difficulties. Tourism came to a standstill and fifty employees were drafted into military service. Those employees were not replaced, and so the Zoo had to get by with its remaining staff. In addition, the management tried to provide financial assistance to the families of the active servicemen. The concert season was curtailed due to a lack of military bands, and in 1915 visitor numbers dropped to the level of 1898 – not least because of the loss of "Berlin's drafted youth", as the annual report stated. The situation the following year was exacerbated by bad weather, preventing any recovery. The impact of the war was now also being felt on the Zoo's animal stocks, as unlike in peacetime, it was not possible to make new purchases to replace deceased animals. Instead, the Zoo made use of the "spoils of war" – the only way to describe its bargain purchase of three goats from Serbia with particularly impressive curly horns. By 1916, all of the Zoo's core staff were at the front, and feeding the animals had become a very challenging task. The difficulties only intensified, and in 1917 the Zoo's supply of oats ceased entirely, which impacted on every single department. The board, however, remained staunchly patrio-

February 1915
Bread is rationed in Berlin

Winter 1916/17
Berlin's population goes hungry during the "Turnip Winter"

August 1917
Food riots in Berlin

Soldiers visit the Zoological Garden, circa 1915

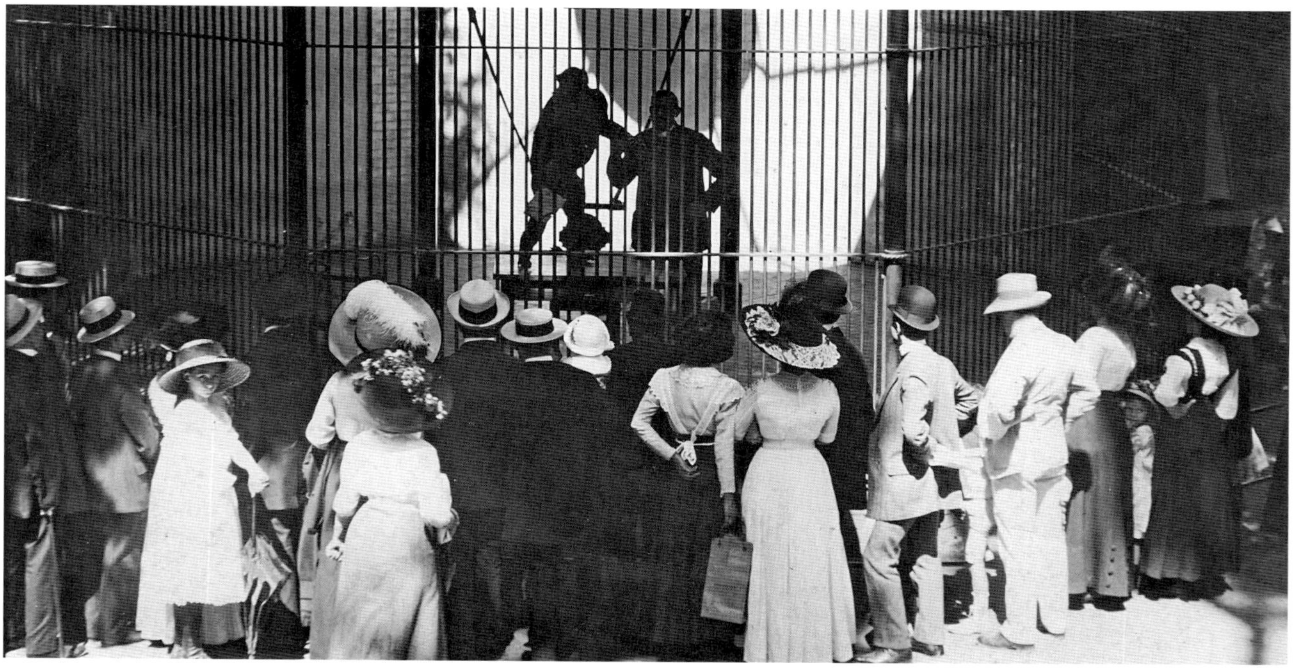

Keeper Liebetreu with a chimpanzee, circa 1915

4 November 1918
The Kiel sailors' mutiny
begins

9 November 1918
Kaiser Wilhelm II
abdicates

9 November 1918
Philipp Scheidemann proclaims
the German Republic

tic, announcing that the Zoo and Aquarium were still better off than the zoological facilities of the enemy. The last year of the war was another bleak one for the Zoo, helped little by the animal gifts that *Reichswehr* soldiers sent from the Eastern Front.[17]

As early as November 1916, the Zoo had to ask the Kaiser for a loan of 200,000 marks. In January 1918, it received another loan of 250,000 marks. Less than a year later there would be no more gifts from his Majesty – in fact, there would be no more monarchy at all. The November Revolution of 1918 brought an end to the German Empire. Once the Kaiser's generals realised that their war was doomed to failure, they left it to the Republican government to conduct the difficult armistice negotiations. Kaiser Wilhelm II abdicated and went into exile in the Netherlands.

The Zoo's debts remained, however. In its annual report, the management board noted succinctly that aid from the emperor was "no longer an option".[18] At the supervisory board meeting on 9 December 1918, just one month after the revolution, no mention seems to have been made of the entirely new situation in Germany. The financial year closed with an adverse balance of 826,000 marks.[19] The Zoo was penniless, with no hope of salvation in sight. Outside the Zoo gates, marauding *Freikorps* paramilitaries ruled the streets. Many of the hundreds of thousands of soldiers now returning home from the front were still loyal to the Kaiser, despite the horrifying war and humiliating defeat for which the monarchy bore much of the responsibility. A large number of these paramilitary groups were radically anti-Semitic and anti-democratic. One of the most infamous *Freikorps* units, the Ehrhardt Brigade, even took up temporary quarters in the Aquarium.[20] In January 1919, the new provisional government under Friedrich Ebert commissioned the paramilitaries to quash the Communists and the Spartacist uprising. One *Freikorps* member who fought in the spring of 1919 was a young lieutenant with experience of the front – Lutz Heck, the son of Zoo director Ludwig Heck. A little more than a decade later, Lutz would succeed his father as director.[21]

Currency crises

The Zoo's seventy-fifth anniversary fell in the midst of those difficult times. Due to the economic crises in the wake of the World War, any hopes of early financial recovery were in vain. In 1920, however, the Zoo received badly needed donations from multiple municipalities – from Spandau to Adlershof, from Wittenau to Mariendorf. This was probably a result of their incorporation into the city, with the formation of Greater Berlin. The donations totalled more than 100,000 marks, with half that sum donated by Charlottenburg alone.[22]

Already in the summer of 1919, however, it became clear that drastic austerity measures would be necessary to prevent the Zoo having to close completely over the winter months. The management board decided to raise admission fees, close some ticket booths, and reduce the still very generous opening hours.[23] And yet all those measures and donations combined were still not enough. Germany had left the gold standard after the war began: the country's leaders had been willing to accept the temporary devaluation of the mark as they were convinced that the German Empire would achieve a swift victory and the defeated nations would have to bear the exorbitant costs of the military confrontation. Coins, which had previously been cast from precious metals like

30 December 1918
Founding of the Communist Party of Germany

January 1919
Uprising by the left-wing revolutionary Spartacist movement

March 1920
Attempted coup by right-wing forces threatens the survival of the Weimar Republic

gold or silver, were now made from cheaper metals such as zinc, iron and aluminium. The paper money that had been in circulation since 1871 was no longer tied to the gold standard. This resulted in skyrocketing inflation. When the victors demanded reparations from Germany at the end of the war, the government simply started up the printing presses. More and more banknotes were put into circulation, and the cycle of monetary devaluation began to spiral out of control. At one point, banknotes were being printed with a nominal value of 100 billion marks. Wages had to be collected with wheelbarrows. At the peak of the hyperinflation in late 1923, one egg cost 320 billion marks.

In any case, the Zoo's shortfall at the end of 1920 had to be compensated with funds from the Reich president's reserves and the municipal government. At the beginning of 1921, Zoo ticket prices increased further.[24]

On 1 October 1922, the Zoo finally closed its gates in order to save on personnel costs over the off-peak season, which not even higher ticket prices could cover.[25] Only the Aquarium remained open – since it had to be heated anyway for the sake of the animals and the building fabric. The people of Berlin flocked to the Zoo on the final weekend before it closed for the winter, with large crowds gathering in front of the enclosures that housed the most popular animals.

Desperate to raise money by any means possible, the Zoo started a fundraising campaign that targeted banks, corporations and private individuals. It also hoped to gain state aid.[26] Despite the rapidly escalating state of national emergency, both the German and Berlin governments made relief payments to the Zoo. The Zoo reopened its gates on 31 March 1923 for the start of the new season. That may have been a little premature, as poor weather kept visitors at bay. The year 1923 turned out to be one of the worst in the Zoo's history. The quality of the feed deteriorated, and some animals died from eating bad hay. In the final months of the year, the Zoo had "easily replaceable animals of high carcass value" slaughtered in order to cover the cost of feed for the more expensive exotic animals.[27]

Salvation for the Zoo and the entire country came when Chancellor Gustav Stresemann pulled the emergency brake, putting a stop to the excessive printing of banknotes; in autumn 1923, he introduced the Rentenmark, a more stable transitional currency. On 30 August 1924, the Reichsmark became the official currency – and money was finally worth something again. Shortly beforehand, the Zoo had reached an agreement with the Prussian Finance Ministry, which was favourably disposed towards the institution. The wartime loans were converted and the Zoo was granted a respite on repayments until 1934, at a low six-per-cent interest rate. That year, a further repayment plan was agreed upon for the debts remaining from the foundation loans provided by the king of Prussia. The Zoo was able to pay off the remaining sum in 1940.[28]

But it was two beneficial business deals in particular that saved the Zoo. First, it gained much-needed funds by selling a strip of land along the edge of Auguste-Viktoria-Platz (today's Breitscheidplatz) by the Kaiser Wilhelm Memorial Church.[29] In addition, the Prussian state allowed the Zoo to lease part of the land given to it for usufruct to two investors. Along with the rental income from the restaurants and the buildings on its southwest edge, those leases enabled the Zoo to continue to develop despite the ongoing crisis, without the need for permanent government subsidies.[30] The future was starting to look a little brighter once more.

1 October 1920
More than 50 municipalities merge to form Greater Berlin

Back on top:
The "Golden Twenties"

The chimpanzees from Tenerife, circa 1920

Ever since war had broken out in 1914, the Zoo had had no access to the international animal trade. Its stocks had been dwindling due to the wartime blockade followed by a lack of convertible currency and then near bankruptcy. In the early 1920s, the Zoo could barely even pay its employees. Investments were therefore out of the question, as was the purchase of new animals. But then the Zoo received an unexpected offer from the Canary Islands. The Prussian Academy of Sciences had previously set up an ape station on Tenerife, populated with chimpanzees sourced from the German colony in today's Cameroon. The financial difficulties of the post-war period meant the academy was now being forced to abandon the station – a bitter loss for the academy, but a great gain for the Zoological Garden in Berlin. Provided it was able to cover transport costs, the Zoo could receive five adult female chimps and one male. Despite its financial constraints, this was an opportunity the Zoo could not pass up on. In 1920, after a journey of three weeks, the six chimps arrived at their new home in Berlin.[31] There was an initial moment of success when one of the chimps gave birth to the first baby ape born in Europe. Unfortunately, however, the mother died shortly after the birth. It wasn't long before the other females also died, of a bacterial infection. All that remained were the adult male Sultan and the infant. Sultan died of tonsillitis in 1923.[32]

Once the currency had stabilised, things started to pick up and the Zoo began acquiring animals again – from Hagenbeck and the other major German animal trading companies Ruhe and Fockelmann. In July 1924, the Zoo presented some animals that had arrived on a particularly large shipment from Ethiopia. They included leopards, a jackal, hyenas, many smaller predators, porcupines, ostriches, aardvarks, giant tortoises, antelopes and more than 1,700 birds. The special attraction was a large troop of hamadryas baboons. The Zoo had been granted a kind of right of first refusal for the animals captured on this large expedition because it had loaned highly experienced zookeeper Petrus Olesen as an expert.[33]

Olesen went on additional expeditions the following year and in 1927/28 – this time with Lutz Heck, the assistant to and son of Zoo director Ludwig Heck. Those expeditions brought in a large number of new animals and enabled the Berlin Zoological Garden to slowly but surely regain its position as the German zoological facility with the greatest diversity of species. Its new acquisitions from those expeditions included a black rhinoceros, five giraffes, a hippopotamus, and a number of predators. A large crowd gathered in the pouring rain at Anhalter Bahnhof railway station to get a first look at the exotic new animals and their African attendants.[34]

In 1928, the Zoo in Berlin achieved the first

20 June 1920
First elections to the new city council

11 October 1920
Official founding of Greater Berlin

The Ape Conservatory, circa 1930

successful breeding of captive orangutans. The infant's parents had been acquired only two years earlier and were a major attraction. Unfortunately, the baby orangutan died not long after it was born. One highlight in the Zoo's recovery phase was undoubtedly its first successful elephant breeding attempt. In late summer 1928, Asian elephant Toni II gave birth to a female calf named Kalifa. As always with such an event, Berliners flocked to the Zoo to admire the new baby. The crowds grew so large that the police had to intervene to "regulate the traffic". The only other time in its history that the Zoo has experienced such a response from animal fans was during "Knut mania" in 2006, when a hand-reared newborn polar bear won the hearts of people around the world.

Elephant mother Toni II took exemplary care of her offspring, and Kalifa grew up perfectly healthy. Sadly, however, she died eleven years later from injuries sustained from spikes on the perimeter of her outdoor enclosure.[35] But Kalifa's birth was not the biggest event of 1928, as it was eclipsed by the arrival of a gorilla, Bobby. The charismatic great ape went on to achieve fame as a very important character in Zoo Berlin's history – even eventually becoming immortalised in its logo. Bobby came to the Zoo at the age of four,

24 June 1922
Assassination of Reich Foreign Minister Walther Rathenau

1 October 1922
Berlin's waste collection is taken over by municipal joint-stock company BEMAG

1924
Germany's first traffic lights go into operation on Potsdamer Platz

Mother elephant Toni II and her calf Kalifa, 1928

Visitors crowd around the Elephant Pagoda after the birth of Kalifa, 1928

Interior view of the Ape Conservatory, circa 1930

along with a chimpanzee named Bubi. The two apes were provided by Hanover-based company Ruhe.

In general, the "Golden Twenties" were the decade of primates at the Zoological Garden in Berlin. The opening of the Ape Conservatory in 1925/26 was a visible sign of the Zoo's recovery following the years of war and inflation. The Ape Conservatory was so large and attractive that it was nicknamed "the palace of the apes".[36] This new animal house had so much space that the Zoo could now show a family group of each species in separate cages. That put an end to fights between different types of apes, which had previously been a frequent occurrence. The plan was also to move the large group of baboons to the southern edge of the enclosure once the Monkey Rock was completed, creating even more space inside the house. At the opening ceremony of the

30 August 1924
Introduction of the Reichsmark as the official currency

24 November 1924
Electricity company BEWAG begins operating electrical facilities on behalf of the city government

Bobby as an infant, 1928

Ape Conservatory, Ludwig Heck claimed that the structure was the only one of its kind in Europe. It also heralded in a change in animal house construction: from that point on, only bar-free rock-based, or water-based habitats were built.[37]

Going cage-free

Although Carl Hagenbeck's plans to create another animal park in Berlin had been thwarted in 1911, his novel idea of displaying exotic animals in a cage-free environment that more closely resembled their natural habitats had put pressure on zoo directors to change their approach. Hagenbeck had not been concerned with presenting as complete a collection as possible of all members of a genus, but with creating an audience attraction where visitors could have an

The Ape Conservatory, circa 1930

1925
Berlin's population reaches more than four million

September 1926
Completion of the Berlin Radio Tower

exciting experience of the natural world. Today, he is largely credited with the invention of zoo animal enclosures without bars, but that is only partially true. Even before Hagenbeck's animal park opened in Stellingen near Hamburg in 1907, zoologists in Germany and Europe had long been developing ideas for cage-free enclosures.

For example, Berlin Zoological Garden's first enclosure with no bars and a natural-looking habitat design opened in 1903 for aquatic rodents. This habitat was created at the southern end of "Lake Lucerne", on top of the large heap of earth that had been excavated to create the lake. Here, homes were established for nutrias, capybaras, beavers and other rodents. The habitat, which was designed to reflect the rodents' homes in the wild with its various pools, was situated lower than the paths surrounding it, creating steep sides that prevented the animals from escaping. Visitors could thus observe the animals from above without any bars obstructing their view. A second bar-free landscaped habitat was created in 1910: the Eagle Rock. Here, the aviaries were grouped around a towering sandstone rock that contained the birds' indoor enclosures. The natural feel of this facility was reinforced by the addition of Alpine plants, mosses and ferns.

In general, however, the managers of the Zoological Garden in Berlin were sceptical about Hagenbeck's approach. Heck still believed that the systematic exhibition of multiple related species and subspecies in neighbouring cages and enclosures was more informative and educational for the public. But he was not able to resist the modern trend. By the late 1930s, the Zoo had fully embraced the concept of cage-free enclosures, adding a large rocky habitat for ibexes and chamois, as well as an open outdoor lion enclosure. The man responsible for this, however, was a much younger zoologist – Ludwig Heck's own son Lutz.

The aquatic rodent habitat, circa 1905

As an assistant to his father, Lutz Heck carefully embarked on a new path, gradually removing bars from enclosures where this could be done "without harm to the animals and danger to humans", ushering in a new era.[38] From the mid-1920s onwards, Heck the younger developed open facilities for elephants, moose, ostriches, bison and zebras. He also had the bars removed from the hippo and camel enclosures, and built the Monkey Rock.

It seems Lutz Heck realised that the Zoo would probably fall behind if it did not make these important adaptations – especially given that the idea of a second, Hagenbeck-style zoo in Berlin reared its head again in 1928. That year, the municipal authorities were repeatedly pushing for further reductions in the Zoo's ticket prices for the lower classes by extending the popular "low-cost Sundays" to also include Saturday afternoons.[39] Meanwhile, some critics were suggesting that the Zoo no longer had enough room for its animals because of the space taken up by the many magnificent animal houses, the wide avenues, and the vast restaurant complex.

A number of prominent Berliners, including writer Alfred Döblin and artist Käthe Kollwitz, founded an association to create and promote a "Volkstierpark" ("people's animal park") and appealed to the public to support the endeavour.[40] Kollwitz was a beloved public figure, and her endorsement carried a lot of weight. On 29 May 1929, she became the first woman to be awarded the *Pour le Mérite für Wissenschaften und Künste* Prussian order of merit, which recognised contributions to the arts and sciences. But the primary initiator behind the new zoo was Clemens Freyer, the recently dismissed administrative director of the Zoological Garden. The planned animal park was to be created in the east of Berlin and was to be much larger. It was intended primarily as a wildlife park, occupied by red deer, fallow deer, roe deer, wild boar and pheasants. However, the backers of the new animal park also wanted to present a few crowd-pleasers such as seals, polar bears and penguins. They also foresaw creating an "African plain" for wildebeest, antelopes and zebras, a "bird meadow" for cranes, flamingos and storks, a large open space for monkeys, and a bear cave. According to the plans, admission prices to the animal park were to be only half that of the Zoo.[41] Initial reactions in the press were positive. Favourable comments came not only from left-wing and liberal newspapers, but also from the conservative press. The *Deutsche Tageszeitung* took the view that the new *Volkstierpark* did not have to be seen as competition to the Zoological Garden, but would actually complement the Zoo as an important local attraction in the eastern part of the city. So much so, in fact, that the paper believed it was clear the two institutions should cooperate.[42]

However, rumours then began to circulate that Hagenbeck's company was sniffing about again and planning to create an animal park in the west of the city, south of Wilhelmstadt and close to the S-Bahn tracks on the major road to Spandau. This new attraction was to be even bigger than the original animal park in Hamburg.[43] The open polar bear enclosures alone would allegedly be big enough to house 120 bears! As if that were not enough, other rumours started that the Zoological Garden itself had plans to relocate to the western outskirts of the city.[44] This would give it more space, the story went, and better opportunities for further development. If the Zoo abandoned its current site, that would also mean wider roads could be created for the traffic in the western city centre, which had risen sharply in recent years. Some motorists were calling for new roads to be built directly through the Zoo.[45] Local politicians in Spandau helped spread the rumours about the Zoo's relocation to the west of the city.[46] Perhaps they wanted to put Ludwig Heck under pressure to move the popular recreational facility to their district. The house and property owners' association in the Berlin district of Schmargendorf, meanwhile, offered the Zoo management the site of its former gasworks, probably in the hope that a new zoo there would attract visitors to the district.[47] The Treptow district office also claimed that it could make several hectares of land available.[48] The site the officials had in mind would have been four times the size of the Zoo's current location.

The Zoo's managers were concerned about a scheme apparently being hatched by their counterparts at the Hagenbeck animal park near Hamburg. Lutz Heck and administrative director Hans Ammon thus set off on a secret mission to the north. The official reason for their visit was to select new animals for the Zoo's outdoor facilities, but they were actually there to finagle information about the Hagenbecks' plans. While Heck examined the animals, Ammon spoke to

10 December 1928
Founding of Berlin transport authority BVG

1929
Alfred Döblin's novel *Berlin Alexanderplatz* is published

Heinrich Hagenbeck, one of the sons of the late Carl Hagenbeck. He assured Ammon that the press reports were untrue. But to be on the safe side, Ammon cautioned Hagenbeck that such a project would seriously damage business relations between the Hagenbecks' animal park and the Zoological Garden in Berlin.[49]

In the meantime, Ludwig Heck began composing numerous newspaper articles advocating for his systematically laid out zoo. In late 1929, he addressed representatives of the press, emphasising the Zoo's animal breeding successes, defending its systematic approach, and highlighting the artistic and architectural value of the Zoo's buildings. His most important argument, however, was the immense costs that constructing and maintaining such a large animal park would entail – ultimately leading to a need for the same ticket prices as the existing zoo. He also claimed that the city of Berlin did not have enough visitors to keep two zoological facilities in business.[50] Since the Zoo also had to accommodate its critics, however, it did offer to help set up enclosures for native fauna in public parks and forests that would cost nothing to visit, thus fostering a closer relationship with nature among broad sections of the population. Such offers had already been taken up in several public spaces: the Viktoriapark and Jungfernheide municipal parks, and the forests of Plänterwald and Grunewald. In addition, the Zoo had released thirty swans on the River Havel.[51] Gradually, the furore died down, and with the onset of the Great Depression the following year, major investments such as those that would have been required for the new animal park became impossible anyway.

Ludwig Heck's son and assistant Lutz Heck continued to plan his renovation of the Zoological Garden, fully aware that it was still in need of modernisation. The first project was the construction of a new rocky outdoor habitat for the seals. Various seal species had been kept at the Zoo since the 1870s. Now, in 1931, an ultra-modern ensemble of various pools and spectator stands was erected next to the eastern end of the cattle ranch. There was a smaller pool for eared seals, one for elephant seals (at times occupied by a breeding pair), a large pool for sea lions, and a pool for penguins. Visitors could look down on

Part of the seals' outdoor habitat, circa 1935

the animals from a rocky overpass, and watch the sea lions swimming in an additional pool through windows set into a rock tunnel.

This new facility set standards for its day, anticipating many concepts that are often found in zoos today, such as the possibility of observing animals underwater through glass. The seal habitat was populated with numerous seal and sea lion species, some of which were kept together despite originating from different parts of the globe. The same was true of the penguins. The highlight of the penguin habitat would undoubtedly end up being the emperor penguins that were presented to the Zoo by Hermann Göring in 1939. These penguins had been brought back

Hamadryas baboons at the Monkey Rock, circa 1930

that same year by researchers returning from the German Antarctic Expedition and gifted to Field Marshal Göring, who had ordered the expedition. One purpose of that expedition was to pave the way for future territorial claims on parts of Antarctica. Sadly, the highly sensitive emperor penguins passed away one by one over the next few years, but Göring's gift was a sign of the good relations to come between himself and the Zoological Garden in Berlin. The original seal and penguin complex is largely preserved to this day.

The next major project was the construction of the Monkey Rock on the site of the old monkey house. This was finished in 1932, to a design based on observations Heck had made during his expeditions in Ethiopia. Other zoos already had open enclosures for their primates. The one now created in Berlin featured a cliff of natural Elbe sandstone with the animal house built into the bottom. Here, visitors could see into the indoor enclosures. The larger habitat on the eastern side of the rock was intended for the hamadryas baboons, while the smaller one on the western side housed gibbons and spider monkeys. The former was taller and rockier to resemble the baboons' natural home in the wild, while the latter was flatter and equipped with trees and ropes for climbing. At the same time, larger bar-free outdoor areas were added to the Elephant Pagoda to provide the Zoo's largest inhabitants with more space.

The Zoological Garden organised its first children's zoo in the summer of 1931, and this quickly became a popular attraction. Here, children could feed and stroke the young of native domestic animals and even the cubs of exotic predators like lions and bears. On the one hand, the Zoo management introduced this new attraction to inspire children with a fascination for animals;

October 1930
Opening of the Pergamon Museum

31 July 1932
The Nazis emerge as the strongest party in the Reichstag elections

The children's zoo: young visitors play with bear cubs, circa 1931 (top); a young elephant and chimpanzee, circa 1932 (bottom)

Indian elephant Harry in his bar-free outdoor enclosure, circa 1933

on the other hand, they knew it would probably lead to an increase in visitor numbers and thus revenue. In a way, the children's zoo served as a substitute for the now less popular and no longer profitable "ethnological exhibitions" of human beings.[52] Above all, however, it enabled children to get closer to animals and gave city kids in particular a great chance to connect with the natural world. Back then, no one knew or cared that creating these sorts of encounters between humans and predators did not constitute appropriate animal-keeping. The young animals at the children's zoo clearly didn't come from the Zoo in Berlin itself, as it did not have enough of its own. Instead, young animals were provided by trading company Ruhe, which also shared in the proceeds. The children's zoo brought in such good money that it became a permanent feature for a time.[53]

Other German zoological gardens had similar children's zoos. Today, neither Zoo Berlin nor any other zoo in Germany allows visitors to interact with predators in this way. What Zoo Berlin does have, however, is a petting zoo, where children can feed and stroke domestic breeds such as goats, ponies and sheep. The petting zoo is a great way for younger visitors to get an exciting up-close experience of animals and learn about them in a playful way.

6 November 1932
Last free Reichstag elections

In thrall to the Swastika: Zoo Berlin during the Third Reich, 1933–1945

Instant assimilation

Many sections of German society had sympathies for Adolf Hitler's National Socialist German Workers' Party (NSDAP, or Nazi Party) even before it came to power as a coalition partner with other right-wing parties in January 1933, following a chaotic election. Some members of the Zoo Berlin staff counted among the admirers. On May Day 1933, many Zoo employees attended the march organised by the Nazis on the Tempelhofer Feld parade ground, with several of them already donning uniforms of the *Sturmabteilung* (SA) or the National Socialist Industrial Cell Organisation (NSBO). As part of the Nazi process of *Gleichschaltung* (the policy of enforced conformity), most of the Zoo workforce were transformed into loyal followers.

The Zoo's music band were soon to be seen posing next to swastikas and the flag of the German Labour Front (DAF), which the Nazis had created to replace the trades unions. The Zoo employees' service club even had a kind of Nazi propaganda shrine.

Berlin's Zoological Garden – complete with its management board, supervisory board and staff – fell in line with the new regime remarkably quickly. As early as late May 1933, the supervisory board unanimously accepted the management's proposal to reduce admission prices for members of Nazi organisations such as the SA, the SS, and the National Socialist Motor Corps (NSKK), but also for members of right-wing veterans' association *Der Stahlhelm,* as well as for their spouses and children. Until then, the discount had only applied to members of the military.[1] Other German zoos had not yet introduced such discounts for Nazi organisations. The following year, the admission prices were reduced permanently for all visitors "in accordance with the high goals of the National Socialist leadership to make it easier for the people to visit cultural sites".[2] Just four years earlier, the Zoo management had vehemently opposed requests from the city government to introduce more days with discounted admission.

Both the former long-standing Zoo director Ludwig Heck and his son Lutz – who succeeded his father in 1932 – had close contacts with right-wing conservative nationalists. Their political views were informed by their enthusiasm for colonialism and the belief that biology should play a central role in the way a state and society are

European bison bulls Spretan and Berthold at Zoo Berlin, 2016

European bison

The European bison (*Bison bonasus*), or wisent, is the largest land mammal native to Europe. It lives in deciduous and mixed forests, where it feeds on herbaceous plants, grasses, leaves and lichen. European bison live in herds of females and their young, led by a matriarch. Adult males only come into contact with these herds during mating season. Bison sport a hump on their shoulders composed of muscles supported by extended vertebrae. Adult bison can grow up to three metres in length and males can weigh as much as a tonne, while females weigh up to half a tonne.

The European bison's original range covered western, central and southeastern Europe as far as the Volga. Hunting and habitat destruction drove these bison to the brink of extinction by the early twentieth century. In August 1923, zoologists responded to the emergency by founding the Society for the Protection of the European Bison. The goal of this international society was to create a record of all bison kept in game reserves and zoos and initiate a breeding programme. As there were very few surviving wisents, many of which were related, some were cross-bred with American bison in order to achieve greater genetic diversity. The characteristics of the American species were subsequently suppressed through further breeding. The basis for all breeding programmes was a centrally managed studbook, which is believed to be the oldest breed registry for a wild animal species.

Both Zoo and Tierpark Berlin have long been involved in European bison conservation efforts. To date, more than 200 of the animals have been born at Berlin's two zoos. Currently, almost 1,500 of these magnificent animals are living wild in the European Union. Like Alpine ibex, Przewalski's horse and the bearded vulture, the European bison is an example of successful species conservation through international cooperation between zoos. European bison have been kept at Zoo Berlin since 1872, when the Russian tsar visited Berlin and was so impressed that he presented one of the animals to the Zoo. This gift influenced the Russian-style design of the Zoo's wisent enclosure in the early twentieth century. Around 1928, the habitat was extended to provide a more spacious outdoor area.

As will be discussed later, it was primarily European bison that brought the two passionate hunters Lutz Heck and Hermann Göring together, thus strengthening the connection between the Zoo director and the National Socialist regime.

A group of European bison at Zoo Berlin, 2006

Zoo employees rally on May Day, 1933 (left and above)

The Zoo's music band, circa 1935

organised. Although it must be remembered that the Zoological Garden was situated in the capital city of the Third Reich and thus had a special status, the actions of other German zoo directors at that time show that there was room for manoeuvre and the Zoo did not automatically have to become an instrument of National Socialist ideology. But, motivated by opportunism as well as personal conviction, the Hecks were more than willing to put their zoo in the service of the Nazis. In his memoirs composed in 1938, Ludwig Heck wrote that he had probably been a National Socialist even before the term was coined. A weight was lifted off his mind, he said, when Hitler founded a new state built on science as well as "blood and soil". Now the Germans would be in a position to apply their "national and racial German worldview" in the realm of politics, too.[3] In 1938, a zoologist as respected as Ludwig Heck was under no compulsion to write those kinds of things in order be published. Still, there is no concrete evidence that he was actually a member of the NSDAP. He is also not known to have worked actively for the regime – unlike his son Lutz.

On 1 June 1933, the Zoo's new director Lutz Heck became a contributing member of the SS, the Nazi Party's *Schutzstaffel* ("Protection Squadron"), supporting it with monthly payments.

30 January 1933
Hitler is appointed Reich chancellor

27 February 1933
The Reichstag fire

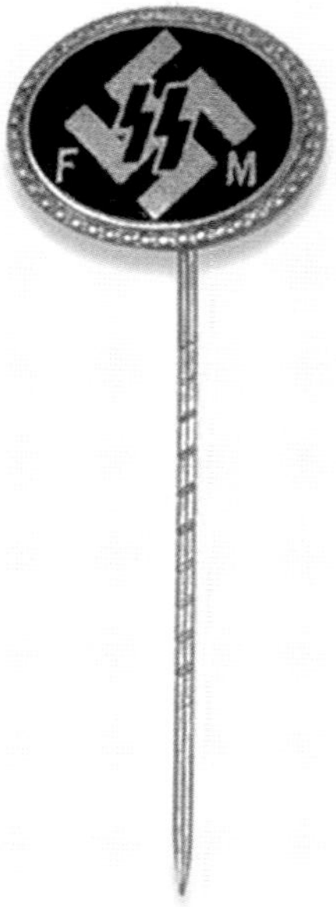

Lapel badge worn by contributing members of the Schutzstaffel (SS)

He identified with the racist and anti-Semitic goals of the SS, but he also wanted to court the favour of the new rulers by offering financial and moral support. The membership badge featuring the two "victory runes" of the SS signified active approval of the regime.

Heck was also a personal friend of Prussian prime minister and Reich Aviation Minister Hermann Göring, who was second in command in the new "Third Reich" after Hitler.[4] Göring was also named Reich Chief of Forestry and Reich Master of the Hunt in 1934. Since their first encounter in 1933 on the Prussian state hunting ground Hubertusstock in the Schorfheide heathland north of Berlin, Göring and Heck had been hatching plans to reintroduce the European bison, partly

Hermann Göring (right) and Lutz Heck at the Schorfheide hunting estate, circa 1935

to satisfy Göring's love of hunting. On that occasion, Heck offered to serve the regime in any way he could. His books and publications praised the wildlife conservation laws passed by the Nazis.[5] At Heck's instigation, Göring was appointed head of a club for breeding hunting dogs.[6] This was certainly not the most prestigious position Göring held, but being given this role no doubt flattered the passionate hunter and won his favour. Heck even provided Göring with young lions as pets, which he returned when they outgrew "petting age". Göring was also gifted a lion from Zoo Leipzig at Heck's behest.[7]

Around the same time, Heck began intensifying Zoo Berlin's bison breeding programme – at Göring's request. In January 1934, Göring presented his plans for the Schorfheide to a group of forestry officials. He wanted to set up a large nature reserve there, where he could also hunt Europe's largest land mammal. Perhaps he pictured himself as a modern-day Siegfried from *The Song of the Nibelungs,* slaying mighty beasts. As ever, Göring's wish was Heck's command, and so he decided to multiply the Zoo's European bison herd as quickly as possible by cross-breeding them with their closest relative, the North American bison.[8] Heck also undertook an expedition to Canada in 1935 "on behalf of Reich Chief of Forestry and Master of the Hunt Hermann Göring, the German hunting community, and the Zoological Garden".[9]

Heck saw National Socialist ideology as perfectly compatible with his profession. In 1941, he wrote that Berlin's collection of animals from all over the world should impress on "the Germans of today" concepts such as the "laws of natural science", an assumed "diversity of human races", and the "destiny of peoples".[10] He was also a man of great ambition with a "thirst for power", as Katharina Heinroth described him after the war.[11] Those traits were particularly noticeable in the way Heck actively sought Göring's friendship. But his own actions as director of Zoo Berlin during the Third Reich also demonstrate that he did indeed share much of Göring's racist ideology. One particularly dark moment in Zoo Berlin's history was the exclusion of Jewish supervisory board members.

Expulsion of Jewish board members

The Zoo's 1933 annual report succinctly notes that the two supervisory board members Georg Siegmann and Walter Simon had resigned "at their own request".[12] That brief summary does not tell the whole story.[13] Before taking the decision to resign, both Siegmann and Simon had been exposed to humiliating discussions at supervisory board meetings – simply because they were Jewish. Siegmann had been a member of the supervisory board since 1928 and its deputy chairman since 1933. He was a lawyer who had represented the Zoo legally for many years. In the first board meeting after the Nazis' "seizure of power", the atmosphere remained collegial and Siegmann was appointed the supervisory board representative in charge of the Zoo's pension fund.[14] Afterwards, however, an off-the-record meeting took place within a closed circle that included Siegmann. It is not known what exactly was discussed at that meeting, but it is likely that the topic was the Zoo's future under the new regime. Contrary to Siegmann's request, no minutes were taken.[15]

At the very next meeting of the supervisory board, supporters of the National Socialists began exhibiting hostility towards their Jewish colleagues. Perhaps to avoid even more trouble,

22 April 1933
Jewish doctors are banned from treating insurance-plan patients

10 May 1933
Book burnings staged by the Nazis take place across the country

Dr Georg Siegmann, before 1938

on 6 April 1933 Siegmann proposed electing Karl Gelpcke as the new chairman of the supervisory board, following the recent death of Bruno Fritsch. Under normal circumstances Siegmann himself would have taken on the position, as the longest-serving board member. The supervisory board approved the proposal, and the minutes note Gelpcke's words of thanks, in particular to Siegmann, whom Gelpcke then nominated as deputy. Normal order was thus maintained, for the time being.

Just minutes later, however, Gelpcke opened a discussion on how to deal with the now vacant seat and the election of new members, which happened on a rotational basis. In view of "today's circumstances", he urged the rapid filling and approval of those positions. Against protests from Siegmann and his colleague Walter Simon, Gelpcke argued that the board must choose candidates "against whom no objections could be raised".[16] What he meant was they should be non-Jews and, if possible, National Socialists. He immediately proposed Duke Adolf Friedrich of Mecklenburg and Wilhelm Hinz. The former had been the last governor of the German colony of Togo, and the latter was a professor at the veterinary college, and a National Socialist.[17] The supervisory board ultimately approved Gelpcke's nominations. The minutes do not record how many voted for and how many against.

Two days later, Mecklenburg was asked by the new chairman of the supervisory board to accept the nomination in these times of "national rebirth", explicitly because of the "close ties between the great colonial movement and our public-serving Zoo".[18] Lutz Heck had already contacted Mecklenburg before the meeting.

Siegmann must have clearly seen the direction things were going, but he was not willing to surrender to the new system without a fight. At the general assembly on 28 March, just two months after Hitler's appointment as Reich Chancellor, the National Socialists among the shareholders launched an attack against the Jewish supervisory board members. A man named Willi Luther called for an amendment to the articles of association that would exclude Jews from the Zoo's management in the future.[19] Gelpcke willingly supported the idea of holding an extraordinary meeting of the general assembly to re-elect the entire supervisory board, and during the following board meeting he called on all the board members to collectively resign. He pointed out that the state might well interfere if such changes were not made to the board's composition. In response, Siegmann stated that giving up his office without a fight went against his personal sense of honour. As a compromise, he proposed amending the articles of association and electing four to five NSDAP members to a larger supervisory board. Otherwise, he said, the Zoo would be in the hands of just a single group, which would undoubtedly be off-putting for Jewish visitors. Gelpcke warned that it was rather Siegmann's insistence that might have negative consequences for Jewish visitors and shareholders. Only "negotiations" would ensure that Jewish shareholders retained their rights and that Jewish visitors would not be harassed.[20] It is horrifying to observe how openly discrimination against Jews was being discussed among the Zoo's supervisory board members such a short time after Hitler's appointment. Unvarnished reference was being made to the very real possibility of harassment of Jewish visitors and board members, and the active participation of Berlin's Jewish population throughout the highs and lows of the Zoo's almost 100-year history was completely ignored.

The supervisory board now sought information from various Nazi Party departments about how many Jewish members would be "tolerable" from their point of view. According to the head of the NSDAP's main economic policy office, *SA-Gruppenführer* Otto Wagener, they would only be able to accept one Jewish board member in twelve. In view of the Zoo's "special circumstances", he did, however, say that a second Jewish member among those twelve could be "conceded".[21] Those "special circumstances" were the many Jewish shareholders and the long tradition of Jewish support for the Zoological Garden in Berlin. The point was made, however, that any changes should be brought about through "negotiation", not by voting.

The question of the Jewish board members was also a main item on the agenda at the fol-

Members of the supervisory board on a tour of the Zoo on 5 July 1941. Eugen Fischer (4th from left) can be seen next to Lutz Heck (3rd from left).

10 December 1933
Erwin Schrödinger receives the Nobel Prize in Physics and flees Berlin soon afterwards

lowing board meeting on 22 May. In the presence of Georg Siegmann, Walter Simon and Paul von Schwabach (a representative of the bank S. Bleichröder who had Jewish ancestors), an unabashed discussion was held as to whether Jews such as these three men could be allowed to remain on the supervisory board. Banker and board member Georg von Simson, who also had Jewish ancestry, did not attend the meeting. It became clear that the Nazi supporters among the board members had few scruples: their Jewish colleagues had to go. To draw as little attention as possible to the matter, a "voluntary" resignation was preferable. The idea of increasing the overall number of supervisory board members, raised again by Simon, was rejected by Gelpcke as this would have required a vote on an amendment to the articles of association. Finally, Siegmann and Simon decided to consult on the question separately.[22]

At the next meeting, on 13 June, Gelpcke shared the most important outcome of their deliberations: Walter Simon had announced his resignation. Adding insult to injury, Gelpcke thanked Simon for his work and the "sacrifices" he had made for the good of the Zoo.[23] Simon continued to work as a legal advisor for several years. He and his wife were later deported to Riga, where they were murdered.[24] Simon's place on the board was taken by National Socialist Otto Wagener, who was co-opted a member at that meeting, without proper election by the general assembly. Siegmann's doubts about the legality of such a procedure were ignored.

In the meantime, the responsible state commissioner in the Prussian Finance Ministry had learned that the National Socialist shareholders, working with the NSDAP press, the SA and the SS, were planning "drastic action" at the next general assembly if their demands were not met. The newly elected supervisory board member Wilhelm Hinz was agitating behind the scenes, calling for complete Nazification.[25] At the next meeting in September, two representatives of the National Socialist shareholder group were co-opted onto the board: their vociferous leaders Willi Luther (a clerk at the justice ministry) and Fritz Kummetz (a police chief).

Gelpcke also stated that, considering "the changed political circumstances", representation of the Zoo by "non-Aryan lawyers" was no longer in its best interest. He suggested that Georg Siegmann should be replaced by a lawyer who belonged "expressly to the National Socialist movement". And he already had someone in mind: Georg Eschstruth,[26] who had been a member of the NSDAP since 1930 and of the SA since 1931.[27]

In the face of ongoing harassment, Siegmann saw no choice but to resign from the supervisory board. In his resignation letter to Gelpcke, he asked that the contents of the letter be made known to the rest of the supervisory board – a request that Gelpcke did not honour.[28]

The board was more cautious with the Zoo's long-standing banking partner, S. Bleichröder. Its representative Paul von Schwabach was more difficult to exclude as, although of Jewish ancestry, he was an important banker and a Christian. This was also true of Georg von Simson. But after continued hostility, even he announced his resignation in 1936. At that year's general meeting, *SS-Brigadeführer* Ewald von Massow was elected to the board in his place. Wagener had been replaced in 1934 by *SA-Standartenführer* Julius Lippert, who was also the state commissioner for Berlin. The general meeting also elected "racial hygienist" Eugen Fischer to the supervisory board.[29] Fischer, director of the so-called Kaiser Wilhelm Institute of Anthropology, Human Heredity and Eugenics, had produced texts

12 March 1934
Defence Minister Werner von Blomberg excludes Jews from the Reichswehr

legitimising, among other things, the "euthanasia" of people with disabilities.

After the death of von Schwabach in 1938, the supervisory board was not only completely "free of Jews", but also largely Nazified. Georg Siegmann and his wife Helene failed to escape from Germany. They were both taken to Theresienstadt concentration camp on 16 July 1942. Two years later, they were deported to Auschwitz, where, like so many others, they were murdered.

The Zoo and the Olympics

For most Zoo visitors, nothing changed at first when the Nazis came to power in 1933. The Zoo continued, for example, to present zoological highlights – like two new shoebills, purchased from the Ruhe company in 1934. These extraordinary birds, whose wise-looking faces and shoe-shaped beaks are fascinating to behold, became great favourites of the public. The Zoo also used the impressive wading birds in its advertisements. In 1935, the Zoo registered a boom in visitor numbers, probably partly because of the agreement that had been concluded with the new National Socialist leisure organisation *Kraft durch Freude* ("Strength through Joy"), which gave its members discounted admission on Fridays.[30]

Director Lutz Heck, meanwhile, continued to implement his extensive plans for further cage-free outdoor enclosures. In 1934, the large facility for red deer was built. The bars were removed from the Stork House, which from then on was surrounded only by a steep-sided moat. The outdoor area for zebras was expanded in 1935, and another was added for wild horses. One milestone in these endeavours to present animals in a new, more open way was the inauguration of the new lion enclosure. With a floor area of 2,000 square metres, it was the Zoo's largest outdoor facility to date. It opened on 24 July 1936, just in time for the Olympic Games, held in Berlin that year. Half of the Predator House was demolished to allow for construction of the new outdoor enclosure, which consisted of a large sand-covered floor – referred to as a "steppe" – set against a backdrop of 1,200 tonnes of sandstone. The steppe featured three elevated platforms that were connected to the building's steam heating system. That meant the lions had warm, dry surfaces to rest on outdoors even during the winter months. For most of the day, visitors could watch a dozen magnificent lions – or alternately tigers, who also used the outdoor space – lounging on the steppe.

Historical Zoo Berlin poster, 1934

15 September 1935
The anti-Semitic Nuremberg Laws are passed

Construction of the "steppe" for the lions, 1935

Opening of the lions' outdoor enclosure, 1936

EBIFA-Druck. Ruf: F 2 Neukölln 2349

Die Deutsche Arbeitsfront
NS-Gemeinschaft „Kraft durch Freude"
Gau Gross-Berlin

Inhaber dieser Anweisung erhält am

Freitag, dem 3., 10., 17., 24., od. 31. 7. 1936

gegen Nachzahlung von RM 0.40 an den Tageskassen des Zoo eine Eintrittskarte für den

„Kraft durch Freude"-Tag im Zoo

05892

Die Anweisung gilt wahlweise auch für das Aquarium.
Kinder unter 10 Jahren in Begleitung zahlen RM 0,25 Eintrittsgeld.

Nur mit D. A. F.- oder N. S. B.-Ausweis

RM 0,05 – Die Anweisung verliert am 31. 7. 1936 ihre Gültigkeit.
Unrechtmäßiger Handel mit diesen Karten wird strafrechtlich verfolgt.

B.-Nr. 686

Nicht übertragbar!

gez. Adam, Gauwart.

Besuchszeiten: Zoo 7—23 Uhr, Aquarium 9—19 Uhr

„Kraft durch Freude"-Tag im Zoo. Juli 1936.

05892

Admission ticket to Zoo Berlin for a festival organised by the Nazi leisure organisation Kraft durch Freude

The 1936 Summer Olympics were a real boost for the still-young National Socialist regime. Although Berlin had of course been granted the Olympics during the short Weimar Republic, the Games represented a welcome opportunity for the capital of the German Reich to present itself to a sceptical global public. All anti-Jewish propaganda was banned from the city. Naturally, the Zoo was also to be incorporated into the Nazi propaganda machine. The Olympic Village just outside the city was adorned with attractive animals provided by Zoo Berlin – but only "German" animals were to be included. The pond was duly occupied with native ducks, geese and swans. There was even a tame roe doe with her fawn.[31] The Zoo also organised a series of special concerts during the Games: naturally, the opening performance was a "grand military concert". The following evening there were performances of music by the Führer's favourite composer, Richard Wagner.

The 1936 summer season was a very successful one for Zoo Berlin. Thanks to the large number of additional tourists in the city, it was able to set a new record of more than two million tickets sold. Ironically, the year when National Socialist propaganda led to the Zoo's greatest success so far was also the year its workforce became a little more international. To make it easier for the city's foreign guests to enjoy a visit to the Zoo, twenty members of staff took language courses so that they would be better prepared to answer the questions of visitors from abroad.[32]

As ever, those visitors mostly came to admire Zoo Berlin's "big five" at that time: elephants, lions, rhinos, apes, and elephant seals. The Zoo even had an extra-special crowd-pleaser up its sleeve: in 1935, a pregnant Asian elephant named Aida had arrived from the zoo in Hanover. In the spring of 1936, she gave birth to a male calf, whom the Berliners soon gave the rather unusual name "Orje" – a dialect form of the name Georg. Orje can still be found at the Zoo today – in the form of a sculpture by the entrance to the Elephant House, polished to a shine by countless children's hands over the years.

The "German zoo"

Lutz Heck's father Ludwig had already established a "patriotic collection" of native animals back in 1888. But those animals were situated at various spots around the park. As early as 1901, there were plans to unite the collection in a single place. In June 1934, Lutz Heck gave an interview to the *Völkischer Beobachter,* the Nazi Party's newspaper. In it, he justified the racial-biological policies of the National Socialists and praised the new "scientific government", which was "politically implementing" the laws of nature. He also spoke out in favour of the establishment of a "German zoo". Initially, however, he said it was sufficient to mark the enclosures of native animals with the colours black, white and red, and the swastika flag.[33]

In April 1935, Heck approached the Prussian Finance Ministry requesting the addition of over two hectares of land on the northern edge of the Zoo.[34] He mentioned that Göring supported this

Visitors queueing up at the Elephant Pagoda, 1936

17 June 1936
Reichsführer-SS Himmler is named chief of the German police

1 August 1936
Opening of the Summer Olympic Games in Berlin

Map of Zoo Berlin from 1937, with the new "German zoo" enclosures visible on the northern edge

Traditional Low German farmhouse and stables, circa 1938

request. Remarkably, however, representatives of the City of Berlin, and of the cavalry divisions of the SA, SS and *Reichswehr* raised objections. They feared the plan might encroach on the neighbouring riding grounds and the bridlepaths along the riverbank. In the end, however, everyone simply had to accept that they could not go against the wishes of Hermann Göring. After an on-site inspection with all involved parties, the objections were withdrawn. Already that August, the management board was able to report the successful transfer of the plot of land to the Zoo.[35]

Heck was then able to realise his "German zoo" on the new site. Enclosures were created to house exclusively "native animals" alongside the existing enclosures for chamois, bison, seals and

August 1936
US athlete Jesse Owens wins four Olympic gold medals

eagles. The "German zoo" broke with the otherwise carefully preserved systematic structure at the Zoological Garden. Planned as a unified landscape, it was the Zoo's first area organised purely along geographical lines. The existing cattle ranch and the Pig House were incorporated into the new area. Behind the smaller Eagle Rock, Heck installed large enclosures for lynxes, wildcats and foxes. There was also a pool for beavers and otters, and, on the northwestern side, enclosures for capercaillies and birch grouses, and a rocky habitat for bears and wolves.[36] On 14 May 1937, the first highlight of the "German zoo" opened: a barn for domestic animals and a traditional Low German timber-framed farmhouse. This was intended to demonstrate how closely humans and animals supposedly lived together on the "old German homesteads". The fact that practically all these animals were also found in neighbouring countries of the German Reich was conveniently ignored. The enclosures were apparently marked with small swastikas,[37] and the animals were contrasted with more "primitive breeds" from abroad.[38] However, it was still the more exotic animals that graced the covers of the Zoo's printed guides – those were, after all, what visitors were most interested in.

The considerably larger area for brown bears, which opened in autumn 1937, certainly improved conditions for these European predators. Their outdoor enclosures were no longer designed like fortresses; instead, the bears were visible from all sides. Alongside the bear enclosure, a facility was created for another "German predator", the wolf. Both new outdoor enclosures were important features of the "German zoo".

To the south of this complex was the old Chamois Rock, inaugurated in 1899. This was now complemented by a second rocky mound made from artificial stone, which opened in 1939 with ibexes as its inhabitants. At its foot were rockeries with central European mountain flora and additional small outdoor enclosures for marmots, rock partridges and salamanders. A trout stream completed the ensemble. This small

Outdoor bear enclosure

"mountain" was cleverly designed: the enclosures and vegetation in front of it meant that the surrounding trench was not visible to visitors, intensifying the impression that they were watching the animals in a genuine mountain landscape.

On 1 May 1937, Lutz Heck finally became an official member of the NSDAP.[39] Due to the high number of opportunists rushing to join the Nazi Party after its ascent to power in 1933, the Nazis had imposed a freeze on new memberships from 1 May 1933 to 1 May 1937. Perhaps that was why Heck had initially "only" been a supporting member of the SS. Once the freeze was lifted, membership was only possible for those who had proved themselves during that time. It's perfectly obvious that Heck would have counted among their number. His close ties with leading Nazis and the way he had zealously aligned Zoo Berlin with Nazi propaganda consummately qualified him as a new member. On several occasions in the post-war period, Heck claimed that he had only ever been on the waiting list for Party membership.[40] That assertion is directly contradicted by what he noted on a form for admission to the Reich Chamber of Literature in February 1938, where he correctly stated his membership and provided his membership number.[41]

Bringing primeval oxen back from the dead

The biggest sensation at the "German zoo" was the aurochs enclosure. The aurochs is an extinct cattle species, the progenitor of all European domestic cattle breeds. The last one died in the first half of the seventeenth century. Lutz Heck and his brother Heinz – the future director of Munich's Hellabrunn zoo – had decided as long ago as the mid-1920s, when Lutz was still assistant to his father in Berlin, that they were going to try breed

The Ibex Rock, circa 1940

November 1936
Carl von Ossietzky rece ves the Nobel Peace Prize

30 January 1937
Hitler orders a comprehensive remodelling of Berlin

"Aurochs" at Zoo Berlin, circa 1935

the aurochs back. For that purpose, Lutz brought wild cattle from the south of France to Berlin between 1927 and 1931.[42] He and his brother were planning nothing less than the "resurrection" of Europe's primeval oxen. They had different ideas about the traits of the aurochs, and therefore included different domestic and wild cattle breeds in their respective experiments. That meant two different lines of "Heck cattle" ended up being bred. Heinz Heck claimed the first birth of an aurochs-like animal in 1932 at his Hellabrunn zoo.[43]

Shortly after the National Socialists came to power, the Heck brothers gained an important advocate for their project: Lutz Heck was able to get Hermann Göring excited about bringing back the aurochs.[44] From 1934, Heck presented the experiments, which had now been ongoing for about a decade, to professional audiences, and in 1937 he exhibited his "primeval German cattle" in a central area of the International Hunting Exhibition.[45] The name of the exhibition betrayed the real purpose behind the breeding experiments. It was not the forefather of all *European* cattle breeds that he aimed to bring back to life – no, he wanted a *German* original: a strong, formidable and awe-inspiring creature that would stand in

April 1937
Jews are henceforth excluded from PhD programmes

Summer 1937
The Nazis remove modern art from museums

Lutz Heck (left) in the Zoo's utility area standing next to a car presumably provided by the Reich Hunting Office, complete with insignia and driver, circa 1941

stark contrast to the weak, "degenerate" domestic cattle. He wanted a Teutonic beast, which "like our heraldic eagle, should be regarded as a true symbol of German strength and courage".[46] As always, Heck's scientific interests went hand in hand with his National Socialist beliefs.

Now, the "new" animal had to prove itself in the wild. Heck was convinced this was the only way an authentic aurochs could rise again.[47] In the late 1930s, some individual cattle were released into the wild on Göring's two hunting estates of Schorfheide in Brandenburg and Rominten in East Prussia.[48] And, indeed, after spending a year in the wilds of Rominten, the animals had – allegedly – become completely wild once more. Heck was overflowing with nationalistic pride: "The extinct aurochs has been resurrected as a German game animal in the Third Reich."[49] The press took up the topic, always referring to the "new breeding" or "resurrection" of this German original. One of the many results of the breeding attempts – a bull – was served up to Heck's father Ludwig on the occasion of his eightieth birthday.[50]

26 February 1938
Opening of the *Degenerate Art* exhibition in Berlin

13 March 1938
The Wehrmacht invades Austria

In the eyes of the two brothers, the main purpose of the aurochs was to fulfil a nation-building function and supply fresh, "wild" blood to the allegedly degraded domestic cattle breeds.[51] Ultimately, it cannot be overlooked that after 1933 the breeding programmes were primarily motivated by ideology, propaganda, and nationalist populism. Heck repeatedly referred to *The Song of the Nibelungs,* which extols the Germanic hero Siegfried and his aurochs hunts.[52] Only in a fantasy world where "the German man" was imagined as a bold hero living in the woods could a wild aurochs serve a meaningful purpose as "a primeval German game animal". However, neither Lutz nor Heinz Heck was actually able to bring the aurochs back to life. The experiment was, and still is, widely regarded as unscientific, and there are good reasons why other zoos did not participate in the breeding programmes.

On 20 April 1938, on the occasion of his forty-ninth birthday, Adolf Hitler conferred the title of "Professor" on several scientists he deemed worthy, including Lutz Heck. From that point on, Heck, who had never worked at a university in his life, used the title in all his correspondence. He even had people refer to him by the title in the immediate post-war period. In the summer of 1938, Göring, in his role as Minister of Forestry and Master of the Hunt, appointed Heck head of the Reich Forestry Office's nature conservation department.[53] After the war, Heck falsely claimed that he had turned down this post.[54]

Incidentally, "Heck cattle" still exist today. The Berlin line did not survive the war, but there are living descendants of the Munich "aurochs" bred by Heinz Heck. In 1980, Tierpark Berlin's director Heinrich Dathe set up an international studbook for this aurochs-like cattle. He had already acquired some of the animals for the Tierpark in 1956 as he did not want "one of the most remarkable zoological ventures of the past decades" to remain entirely "unsung". He did not, of course, refer to the animals as aurochs, but as "Aurox-like domestic cattle".[55]

"Jews not welcome"

When the Nazis came to power in 1933, many companies and leisure facilities throughout the German Reich started displaying anti-Semitic signs. Increasingly, the notice "Jews not welcome" appeared outside pubs and restaurants, recreational facilities, public baths and even at the entrances to some towns. Such signs did not initially go on display at zoos, however – perhaps out of consideration for international visitors. At Zoo Berlin, the supervisory board started considering excluding Jewish visitors by 1938 at the latest. In March 1938, administrative director Hans Ammon asked his colleagues in Düsseldorf and Wuppertal-Elberfeld how they were dealing with Jewish visitors. It is likely that other zoos also received this query. In Düsseldorf, Ammon's letter was passed on to the Deutscher Gemeindetag (the confederation of German municipalities), which approved discriminatory measures but advised caution. It is not known what answer was received from Wuppertal, but the management there did ask the Zoo Berlin management about possible applicants for animal-keeping jobs, who should be "of German-blooded descent".[56]

At the same time, Ammon tried to find out from the deputy Gauleiter for Berlin how far the Zoo could take "such measures". As a result, the "contemplated measures" had to be postponed, because the Gauleiter's office stated that any discussion of the matter in the general meeting should be avoided.[57] By the end of the year, however, all scruples had been overcome.

26 April 1938
Göring makes it mandatory for Jews to register their assets

4 May 1938
Carl von Ossietzky dies as a result of his brutal treatment in a concentration camp

At a supervisory board meeting on 8 November 1938, Ammon stressed that the "Jewish question" had become a "matter of particular concern" for himself and Director Heck. The chairman of the supervisory board, Karl Gelpcke, said that the time had come for the Zoo to adopt the kinds of signs that had long been on display in public parks. His idea was to start by putting up notices next to the Zoo's nativity scene stating that Jews were not welcome there. Then, from 1 January 1939, such signs should also go on display at the Zoo gates. His idea received the backing of *SS-Brigadeführer* Ewald von Massow and Berlin's state commissioner Julius Lippert. The former also considered it important to create a separate playground for Jewish children. The latter said that if Jews were now not welcome in parks and indoor pools, they should also be denied entry to the Zoo. The supervisory board unanimously decided that season passes should no longer be issued to Jews. The Zoo did not offer any compensation to Jewish holders of season passes. And for Jewish shareholders, it meant they no longer received the "dividends" from their shares – i.e. free admission. When Ammon pointed out that this could lead to Jewish shareholders demanding the Zoo buy back their shares at the nominal price, legal advisor Georg Eschstruth replied that this would probably not happen. Gelpcke also argued that the signs did not constitute an actual ban, they simply showed Jews that they were not welcome.[58]

In any case, the course of history was soon to catch up with the Zoo. On the night of 9 to 10 November, the Nazis instigated a nationwide pogrom against Jewish institutions and businesses. Thousands of livelihoods were destroyed, and hundreds of people were killed or injured. Just one day later, the Berlin police president banned Jews from visiting places of public entertainment. That demonstrates that the measures being debated by the Zoo management were entirely in line with National Socialist policy. However, the management at Zoo Berlin had already been discussing concrete anti-Semitic measures before it had even become official policy in the city, and they were ready and willing to receive complaints from some season pass holders. After the November pogrom, the supervisory board began referring explicitly to a "ban on Jews".[59]

There is evidence that in 1939 a small sign reading "Jews not welcome" hung from the Elephant Gate for everyone to see.[60] At the shareholders' annual assembly on 28 April 1939, the administrative director, standing beneath a picture of the Führer, humiliated the unwanted Jewish visitors even further by claiming that the Zoo had always been "an important source of clean, fresh air for the city of Berlin – and since November 1938, I think, the air here has become even cleaner."[61]

Jews were no longer welcome as shareholders either. The Nazis' increasing disenfranchisement of German Jews robbed them of their social standing, their income, and their future. From 1933 onwards, Jewish people were forced to sell their personal belongings in order to emigrate and, ultimately, survive. Those belongings included Zoo shares, which had often been in the family for generations.[62] These were by no means speculative securities; they had an emotional value for Jewish shareholders in particular, as they were proof of their integration into Berlin bourgeois society. Shares in the Berlin Zoological Garden were never about making money, instead they were usually an expression of patronage and community interest – especially for Jewish shareholders. Purchasing a share was an act of support for the Zoo, helping it to raise urgently required funds. The only real "dividend" was free

28 – 30 September 1938
The Munich Conference authorises the annexation of parts of Czechoslovakia to the German Reich

October 1938
The passports of German Jews are stamped with a "J"

admission for shareholders and their families. Of course, the price of Zoo shares also fluctuated over the years, and sales of those shares could incur a profit or a loss.

By autumn 1938, most of the shares owned by Jews had been sold on. Some who sold early had still found buyers willing to pay the going rate – which had, however, been falling rapidly since the late 1920s. After the November pogrom of 1938, the remaining Jewish shareholders finally tried to offload their shares. And on 3 December 1938, the Nazis gave all Jews just one week to deposit all their stocks in the bank. That meant they no longer had access to their shares. The sudden renewed availability of Zoo shares on the market caused an additional price drop and thus further financial loss for sellers.

Since the shareholder association's founding in 1845 as the Actien-Verein, its articles of association had not provided for any scrutiny of share sales by the management. There was no legal control over who bought the shares, nor did the Zoo have to agree to a sale. The management thus had no means of controlling whether or not Jews were continuing to buy its shares. In the spring of 1938, the supervisory board was therefore making plans to amend the articles of association accordingly at an upcoming general meeting. Instead of the names of new owners simply being entered on the register after the sale, the Zoo would now receive the right to approve a sale or not. There is no reason to doubt that this step was primarily taken to control the sale of shares to Jews. This way, the Zoo could put pressure on sellers to part with their shares for a low price. It is questionable, however, whether the Zoo was actually in the position to buy a large quantity of its own shares. For legal reasons, the supervisory board ultimately refrained from including such a proposal on the agenda of the annual general meeting on 29 March 1938. The restriction on sales would also have affected non-Jewish shareholders, who might have opposed it:

"The proposal of the management board to amend Section 3 to make conveyance of shares subject to the approval of the company in order to gradually eliminate non-Aryan shareholders is, in the opinion of our legal counsel, unfortunately impracticable, because the inclusion of such a provision is only permissible for newly established companies, while for already existing companies the approval of all affected shareholders is required." (AZGB archive O 0/2/2)

The available documentation suggests that some Jewish shareholders who were forced by the regime to sell their Zoo shares contacted the Zoo management requesting them to find buyers. As the sale of the shares of Jewish shareholders to non-Jews was not moving fast enough for the Zoo management, on 19 July 1938 the supervisory board decided to acquire the legally permitted ten per cent of its own shares from Jewish shareholders. The law only allowed for this in order to avert damage to a joint-stock company. "The supervisory board and management board are in agreement that the fact that a portion of the Zoo's shares is still in Jewish hands constitutes serious damage to the company [...]."[63]

The Zoo subsequently acquired around a hundred shares from Jews, which it then passed on to "Aryans".[64] This corresponded to around 2.5 per cent of the shares. The few documented intermediate sales from those months show that the Zoo tried to at least turn a profit on those transactions by buying at a low price and selling at a higher price. One Elisabeth Wartenberg, for example, sold her share to the Zoo on 6 December 1938 for 300 Reichsmarks. The actual price of the share was about 50 marks more. The

9/10 November 1938
The Nazis initiate pogroms against Jews across the country

15 November 1938
Jews are banned from attending state schools

3 December 1938
Driving licences held by Jews are declared invalid

Zoo sold that share just ten days later, probably for 370 Reichsmarks, and kept the difference. The Zoo also ensured that new buyers were not Jewish, sometimes even going so far as to have this confirmed in writing. Wartenberg's share had most likely been owned by Jewish Berliners since its issuance in 1871, as is revealed by surviving Zoo tickets belonging to shareholders of the Gerson family. Wartenberg was deported to the Theresienstadt concentration camp in January 1943, where she died a few months later.

In 1940, the share price was double what it had been in October 1938. At the annual general meeting in May, administrative director Ammon stated that the "reallocation of Zoo shares can essentially be regarded as complete".[65] In 1941, the Zoo bought a batch of its own shares, which had "conspicuously accumulated in a single hand", in order to subsequently sell them individually.[66] Bank clerks had probably been using the shares deposited by their Jewish customers to line their own pockets. The Nazi state also benefited from the sale of Zoo shares. With the introduction of the 11th Decree of the Reich Citizenship Law in 1941, all the property of German Jews who had been deported to concentration camps in the east was confiscated. Their shares were taken over by the Prussian State Bank. The bank then sold those looted shares, offering them at least once to the Zoo at the going rate.[67]

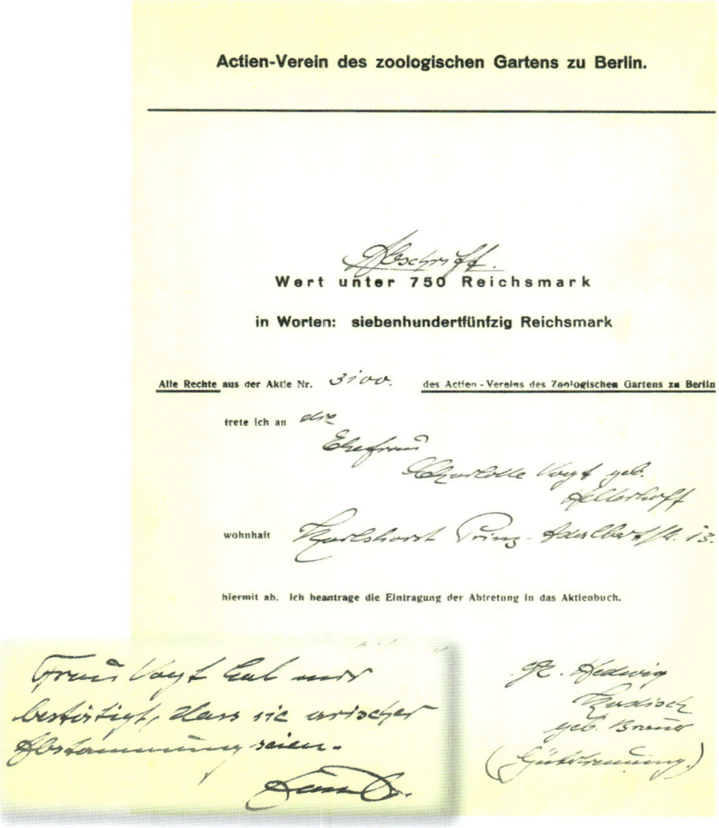

Actien-Verein des zoologischen Gartens zu Berlin.

Abschrift

Wert unter 750 Reichsmark

in Worten: siebenhundertfünfzig Reichsmark

Alle Rechte aus der Aktie Nr. 3100. des Actien-Vereins des Zoologischen Gartens zu Berlin

trete ich an die Ehefrau ... ab.

wohnhaft ...

hiermit ab. Ich beantrage die Eintragung der Abtretung in das Aktienbuch.

Deed of assignment for one share. In the bottom-left margin the new owner confirms that she is »of Aryan descent".

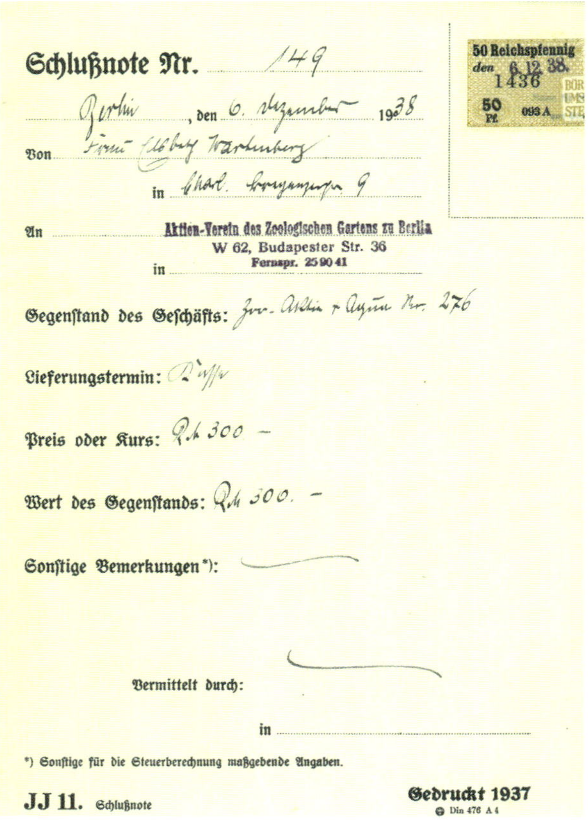

Schlußnote Nr. 149

Berlin, den 6. Dezember 1938

Von Frau Elisabeth Wartenberg

in ...

An Aktien-Verein des Zoologischen Gartens zu Berlin W 62, Budapester Str. 36 Fernspr. 25 90 41

in

Gegenstand des Geschäfts: ... Nr. 276

Lieferungstermin:

Preis oder Kurs: RM 300 –

Wert des Gegenstands: RM 300. –

Sonstige Bemerkungen*):

Vermittelt durch:

in

*) Sonstige für die Steuerberechnung maßgebende Angaben.

JJ 11. Schlußnote

Gedruckt 1937

Assignation note for the share owned by Elisabeth Wartenberg, December 1938

November 1938
Several laws and ordinances are passed that massively impede the economic activity of Jewish people

The Zoo at war again

On 1 September 1939, Nazi Germany launched its pre-planned war by invading Poland. The outbreak of war had massive implications for all countries involved, including their zoos. Employees were drafted into the armed forces, food was rationed, and some people started asking whether a zoo was not a luxury that should probably be dispensed with during wartime. In Germany, that discussion quickly fell silent on 14 September 1939 when Hermann Göring, in his function as head of the Reich Forestry Office, decreed that zoological gardens should remain open during the war, just like museums and theatres, in order to fulfil their cultural education remit. The order was signed by Lutz Heck himself, as head of the Reich Forestry Office's nature conservation department.[68]

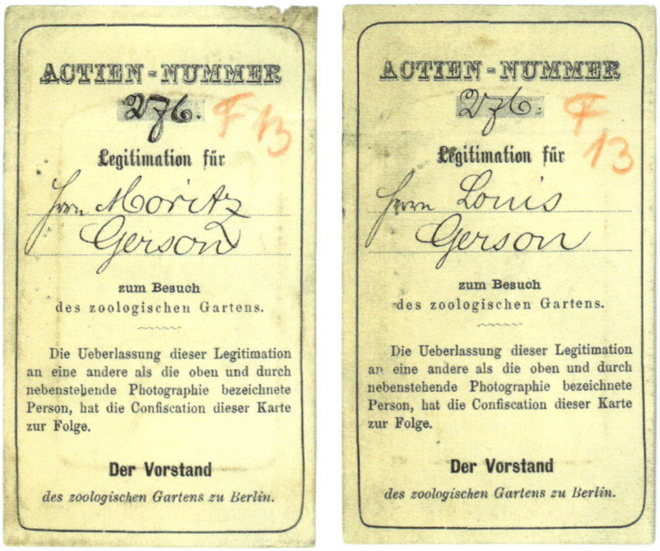

Shareholder cards for share No. 276 owned by the Gerson family, circa 1880

In the first year of the war, Zoo Berlin was spared any war damage. However, all the glass roofs had to be painted with a camouflage coating to prevent reflections. That made the interior of the buildings much darker, of course. But a greater problem for the Zoo was the requisition of the large restaurant ensemble, which was then used to accommodate soldiers. The Zoo lost a lucrative source of income from the lease of those buildings but was at least reimbursed with a lump sum from the High Command of the Wehrmacht. That was important, because the Zoo still could not survive on its own income alone.[69] After some negotiations, the vestibule of the magnificent *Kaisersaal* remained available as an eatery for Zoo visitors and as a venue for the Zoo's annual general meetings.[70]

Inevitably, war led to the rationing of food and animal feed, and that certainly posed some problems for the Zoo. Initially, the issue was resolved thanks to the good will of the authorities, no doubt due to Heck's influence within Göring's Reich Forestry Office. However, the non-exotic inhabitants of the domestic barn were deemed "unproductive" and were slaughtered to "feed the populace". Some ibexes and lynxes were transported to Göring's Rominten hunting estate in East Prussia.[71] Initially, Zoo Berlin had taken in animals from zoos in the western part of the Reich, for example Cologne, Münster, Düsseldorf and Duisburg, as these seemed to be at greater risk from the fighting.[72]

Several zoo animals were even "conscripted" for use by the Wehrmacht. Around sixty animals were issued to soldiers as mascots in 1942 alone, including guinea pigs, raccoons, coatis, dogs, dingoes, wild boars, sheep, goats, monkeys, storks, pheasants, parrots, jackdaws, magpies and jays. These animal companions were intended to provide distraction and amusement for combat units and naval and submarine crews. Conversely, the Wehrmacht also brought new animals to the Zoo from the various war zones.

15 March 1939
Occupation of Czechoslovakia by the Wehrmacht

1 September 1939
Germany invades Poland

3 September 1939
France, Great Britain and the Commonwealth enter WWII

Keeper Liebetreu being congratulated for his years of service by Lutz Heck (right), who is wearing Nazi Party badges, 1940

This is how the Zoo received wolves from Minsk, a bear from Lapland, and a dorcas gazelle from Africa. Wehrmacht units also donated miniature donkeys and a stork to Zoo Berlin.[73] On the home front, the Zoo was primarily dedicated to looking after Wehrmacht soldiers, who were given free admission when they were home on leave as part of the "Strength through Joy" programme or if they were receiving medical treatment.

Everywhere, buildings were being made bombproof and air-raid shelters were being built – and the Zoo was no exception. Shelters intended for Zoo staff were created in two feed storage cellars in the administration and utility facilities on today's Budapester Strasse.[74] An air-raid shelter was also constructed close to the Zoo entrance, under the promenade leading from the Elephant Gate into the Zoo. This was also intended for residents of neighbouring buildings, audiences at the nearby theatre, and guests of the Eden Hotel. The two front rooms were for use by the public while the two rooms at the rear were reserved for Zoo staff. The shelter was shielded by a layer of concrete almost two metres thick and camouflaged from above with a flowerbed.[75] It features on the 1940 map of the Zoo.

The spacious indoor animal accommodation inside the recently built Ibex Rock was also developed into an air-raid shelter. The vaulted roof was 25 centimetres thick and topped by about 60 centimetres of rock. The facility was intended to provide protection for 150 people in the event of a bomb attack, but it seems that ultimately it was never used as an air-raid shelter for humans.[76]

Some zoos in western Germany, Europe and Germany's ally Japan killed their dangerous animals for safety reasons at the beginning of the war or as it progressed.[77] This did not happen in Berlin, however. The lions were considered tame, while the leopards were housed in a very solid building. A guard was assigned to the elephants, and a directory was created of which venomous snakes had to be evacuated to secure places when the air-raid siren sounded. Brick "molehills" were created as ad-hoc shelters for the guards, and for the runners who traversed the grounds on bikes.[78] In 1942, three splinter-proof iron observation booths were built, intended for the detection and reporting of any escaped animals during air raids. The Zoo also established its own fire brigade equipped with a portable motor-

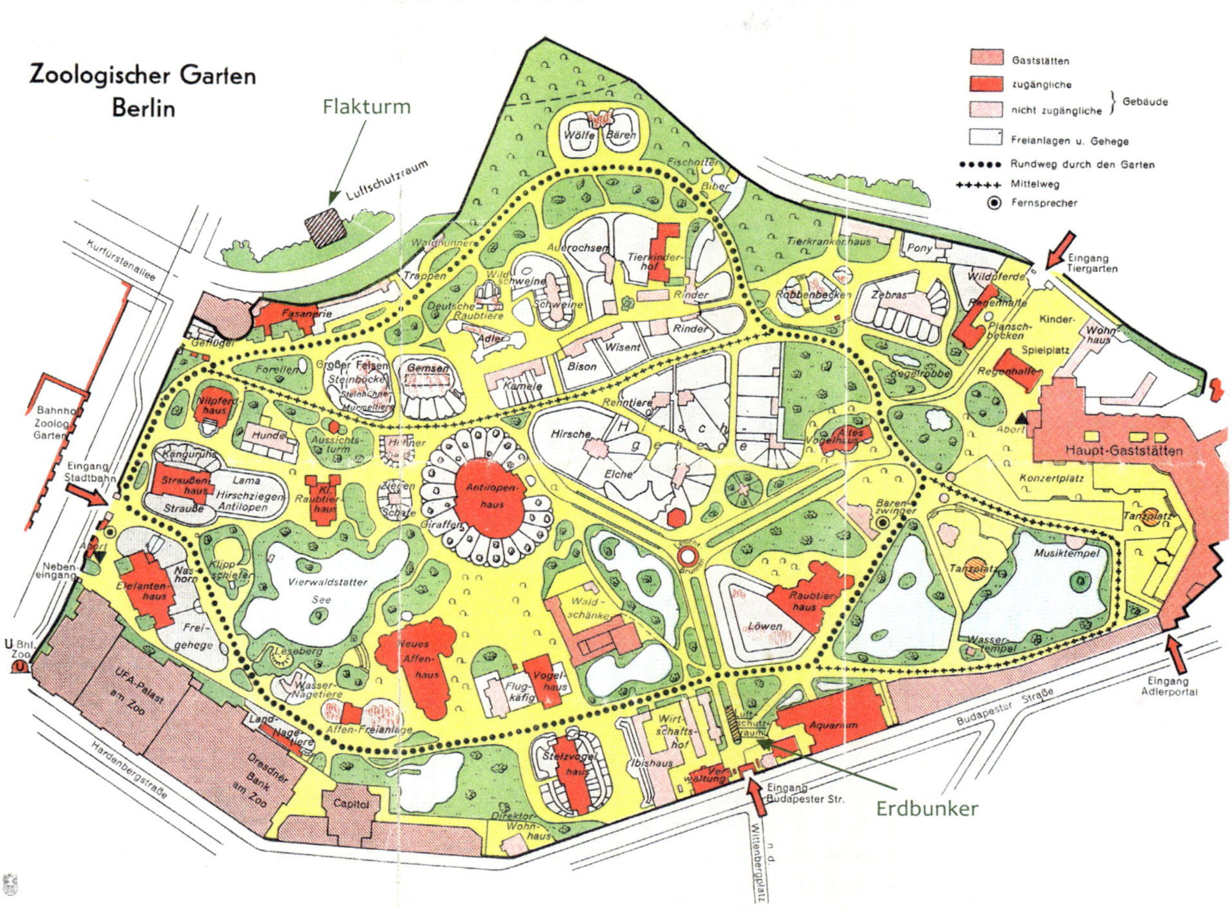

Zoo Berlin map from the 1940 printed guide. The underground bunker (Erdbunker) just inside the Elephant Gate is labelled, as is the flak tower (Flakturm) in the northwest.

ised pump and animal capture teams armed with nets, snare poles and searchlights. Some members of the team even had rifles.[79] For the protection of its visitors, the Zoo was also connected by telephone to the air defence communication network.[80] All these measures had been agreed with the two authorities led by Heck's crony Hermann Göring – the Reich Ministry of Aviation and the Reich Forestry Office. Eventually these measures were installed at all German zoos.[81] They prevented the need for the mass killing of zoo animals until ground combat closed in on zoos in 1945.

In April 1941, a large, square flak tower was built at Zoo Berlin, close to the railway lines. It was 40 metres tall, 75 metres wide, and had concrete walls two and a half metres thick. It was also intended to provide shelter for 15,000 people over five storeys and it contained art treasures from Berlin's museums. The Zoo management had protested in vain against the building of the tower. They feared it would make the Zoo and its environs the target of attacks, and that shot-down planes would damage the Zoo. But this time even Heck's good relations with Göring were no help –

August 1940
The first bombs fall on Berlin

22 June 1941
Germany invades the Soviet Union

he was, after all, also commander-in-chief of the Luftwaffe.[82]

Ultimately, the Allied carpet bombers were not concerned with differentiating between military and civilian targets. As mentioned above, initially the Zoo grounds were spared major damage, while many outlying areas of the city were hit hard during the first bombing raids in 1941. The Allies were still building up their bombing capabilities, and at a time when the Luftwaffe was flying out from airfields in German-occupied countries close to its targets and reducing Warsaw, Rotterdam and parts of London to rubble, the reach and efficacy of the British "night bombers" were still fairly limited. As will be seen later, it was in particular the ground battles around the flak tower in its immediate vicinity in April 1945 that really spelled disaster for Zoo Berlin.

Heck and the "eastern expansion"

In Lutz Heck, Zoo Berlin had gained a director who not only had contacts at the highest level of the Nazi elite, but also wholeheartedly supported their expansionist and colonial ambitions. In April 1933, a replica of the base camp from his African expeditions formed the centre of a colonial art exhibition in the Zoo's exhibition halls.[83] In summer 1934, the Zoo held a special fair to mark "Colony Commemoration Day" and organised a press tour past "German colonial animals". In 1937, the Zoo celebrated a colonial festival under the motto "Everyone should experience Africa".[84]

As head of the Reich Forestry Office's nature conservation department, Heck was responsible for the introduction of the applicable Reich regulations in the occupied Sudetenland and annexed Austria.[85] The Reich Forestry Office later drew up absurd plans for the landscape and nature conservation measures, as well as hunting regulations, that would apply in the conquered African colonies once Germany had achieved its expected "final victory".[86]

The Nazis' expansionist ideas and associated need for *Lebensraum* ("living space") primarily concerned territorial expansion in the east. This was to be achieved through a series of wars in eastern Europe. Heck also had a role to play in the expansion policy. In late September 1939, he received a report from a Dr Müller in Warsaw, which the Wehrmacht had just conquered. Müller gave a detailed description of circumstances at the zoo in the conquered city.[87] The memoirs of Antonina Żabińska, wife of Warsaw Zoological Garden's director Jan Żabiński, record that Heck came to Warsaw himself shortly afterwards and ordered the transfer of the most impressive animals to zoos back home in Germany. Then, at the behest of the Nazi regime, he initiated the liquidation of the Warsaw zoo. The remaining animals were killed on site by SS units during drunken revelries. Director Żabiński and his wife were ordered to grow feed on the grounds of their zoo and set up a pig farm.[88] Heck's report for 1939 stated that he had placed several animals from Warsaw "in trustworthy hands". This was, of course, pure euphemism: in reality, Heck had stolen the valuable animals and divvied them out to German zoos as he saw fit. In the meantime, twenty-eight animals from Warsaw had arrived in Berlin. They included an Indian elephant, two European bison, a wild horse, a wildebeest, a zebra, three camels, an ostrich, four lynxes, seven porcupines, a llama, two Shetland ponies, two flamingos, two pelicans and a black swan. In June 1940, a lioness from Warsaw arrived in Berlin, handed over by a police battalion.[89] And thus Zoo Berlin and other German zoos became war profiteers, while the Nazis decided that Poland, which had been split into two zones following the German and Soviet invasions, no longer had any need of zoos. Poland was to be reduced to the status of a slave nation.

In an odd twist of fate, however, the wrecked Warsaw Zoological Garden was soon to become a place of refuge. Over the following years, the Żabińskis smuggled Jews out of the Warsaw Ghetto and offered some of them shelter in the abandoned zoo. Jan Żabiński later participated in the Warsaw Uprising as a member of the Polish underground resistance. He and Antonina are honoured as "Righteous Among the Nations" at the Yad Vashem Holocaust remembrance centre for their rescue efforts.[90]

Heck also helped himself to other natural resources of conquered Poland. In the first half of 1940, the SS brought a number of Polish "Panje horses" to the Zoo. This was allegedly done in order to preserve the endangered breed. In actuality, Heck intended to cross them with "forest horses".[91] It is evident that his plan had nothing to do with species conservation. Instead, he and his brother Heinz in Munich wanted to use this war booty to try to breed back the extinct tarpan.

Heck ventured even further into the war zone shortly afterwards. The vast tracts of land between Białystok and Brest were home to the ancient Białowieża Forest – the former hunting grounds of the Russian tsars and one of the last primeval forest landscapes in Europe. It also happened to be the last refuge of the European bison. When Germany captured the area following its attack on the Soviet Union on 22 June 1941, Reich

"Evictions" in Białowieża, 1941

Master of the Hunt Hermann Göring suddenly had the opportunity to annex the forest. He immediately established a "Reich hunting ground" in Białowieża. At 260,000 hectares, this was considerably larger than the original protected area. The plans for the hunting ground were, however, hindered by the presence of local people, who lived in small settlements on this land. The "master race" conquerors of the former Polish territory felt that the natives might well cause mischief as potential poachers. And so the decision was made for German "forest protection troops", police battalions and the SS to drive out the local people.

The evictions began without warning, and were swift and brutal. The Germans stole the villagers' livestock, then burned their homes and farms to the ground. Several settlements in the area had a majority Jewish population. During the expulsions, these people were treated with even greater brutality: several hundred Jewish men were killed on the spot, and their families were placed in ghettos, where they were probably subsequently murdered. Heck's superior, Ulrich Scherping, head of the Reich Hunting Office, was in charge of the campaign.[92] Scherping boasted of his "purges" and persecution of Jews in the very same hunting magazine in which Heck published articles on his "aurochs".[93] Heck and his nature conservation department ensured that the hunting and forestry authorities collaborated smoothly with Heinrich Himmler, *Reichsführer-SS* and Reich Commissioner for the Consolidation of German Nationhood in the occupied territories. Heck was therefore almost certainly involved in the campaign, at least indirectly.[94]

For Himmler and for Heck, these areas were not the homes of the people who lived there; they were simply a resource, an empty space that needed to be "civilised" in accordance with their own notions. "The transformation of this desolate foreign landscape into a German one must be our most important goal," wrote Heck in 1942. "For the first time in history, the shaping of a cultural landscape is consciously being undertaken by a people."[95] Heck was on site at least twice between 1941 and 1944, probably also to reintroduce some of his "aurochs".[96] The Zoo's animal register records the arrival of more "Panje horses" in spring 1942, this time from Białowieża, which were part of a larger transport from the area to Munich via Berlin. After the war ended, Polish representatives, with the help of the British occupying forces, searched for twenty-seven of these "feral domestic horses (or konik)", which Lutz Heck, his brother Heinz, and *Oberstjägermeister* Scherping were believed to have stolen.[97] Heck also spent time in a nature reserve in Crimea during the war, and the Soviet army later accused him of looting animals for his own use there.[98] In fact, at least one pair of Przewalski's horses probably came to Berlin that way.[99] All these animals may have been involved in the Heck brothers' attempts to breed back the tarpan in Berlin and Munich.

"Essential to the war effort": Heck exerts his influence

Armed with the official order that German zoos were to remain open even once the war started, Heck vigorously asserted that the work carried out by Zoo Berlin employees was "essential to the war effort" and thus saved them from being drafted. But he didn't always get his way. In May 1940, around thirty Zoo employees were serving in the Wehrmacht. By May 1941, that figure had increased to forty, and at the end of 1943 around sixty staff members had become soldiers.[100] However, in June 1941, Heck's nature conservation

1 January 1942
The Declaration by United Nations
is signed

department attempted to have the management and the particularly reliable keepers at several zoos registered as "indispensable".[101]

To heighten his influence, as early as 1939 Heck attempted to become general supervisor of all German zoos. He was planning the establishment of a central association or organisation of German zoos under the auspices of the Reich Forestry Office, with the war providing legitimacy for the plan.[102] He intended to lead the organisation as a dictator, according to the Nazi *Führerprinzip* ("leader principle"). Although Heck failed in this attempt, as head of the Reich Forestry Office's nature conservation department, he certainly achieved an unprecedented position of dominance over his zoological colleagues. In February 1941, Heck invited all German zoo directors to Zoo Berlin and the premises of the Reich Forestry Office. On the second day of their meeting, the only item on the agenda was the creation of a nationwide association of zoological gardens. Other zoo directors and representatives, as well as various municipal authorities, had raised concerns about the idea of a standardised, homogenous organisation of all zoos. In an attempt to dispel their doubts, Heck presented a statute drafted by the Reich Forestry Office. No one present dared to openly object to this clear attempt on Heck's part to assume dominance over all zoos.[103] Ultimately, however, nothing came of the plan and no such association was founded.

Lutz Heck gives Hermann Göring and his daughter a tour of Zoo Berlin, 19 March 1942

20 January 1942
The "Final Solution to the Jewish Question" is formulated at the Wannsee Conference

February 1942
Fifteen-year-old schoolboys are called up for anti-aircraft service

Soldiers on leave visit the Zoo, 1943

At the Reich Forestry Office, Heck oversaw the supply of feed to German zoos, liaising with zoo directors, suppliers and the relevant authorities. Right up to the spring of 1945, Heck was contacting various zoos to ask about necessary feed quantities for different fish-eating species so that he could coordinate with various authorities and fish wholesalers in organising a share of the by-catch from sea fishing. Thanks to his connections, Heck was able to secure a supply of feed for the few remaining animals for a surprisingly long time. He was particularly concerned with safeguarding the "valuable" seal and penguin stocks. But in the end, of course, this too became impossible. From spring 1945, the *Reichsbahn* rail authorities refused to transport any animal feed not considered essential to the war effort.[104]

But that was only at the very end. Despite the adversities of the war, the Zoo tried to keep visitor numbers up – and at first actually succeeded. In 1940, the Zoo and Aquarium welcomed just as many visitors as in the Olympic year of 1936. The year 1942 was also a good, thanks to excellent ticket sales and increased admission prices. The Zoo's sustained popularity may have been partly due to the many concerts it held, with at least one every day.[105] Like many prominent politicians, Hermann Göring also used the Zoo as his

March 1942
The mass depo-tation of Jews to Auschwitz begins

March 1942
Jews are forced to mark their homes with a yellow star

personal promotional space. In March 1942, for example, he visited the Zoo with a large entourage as well as his three-year-old daughter Edda.

A visit to the Zoo remained a favourite pastime of Berliners during the war. Even after Germany declared war on the Soviet Union, visitor numbers remained stable and high. Administrative director Hans Ammon speculated that this was due to the otherwise limited opportunities for recreation and the "numerous foreigners employed in Berlin". The latter was a veiled reference to the foreign forced labourers who had been put to work throughout the city, including at the Zoo. Between 1940 and 1945, there were tens of thousands of forced labourers in Berlin, most of them from Eastern Europe. Many visited the Zoo in their limited spare time as a welcome distraction. They were banned from visiting most other recreational sites and cultural facilities, and would be punished if they did so. After liberation in 1945, many former forced labourers reminisced about their visits to Zoo Berlin. The photograph shows a Polish forced labourer named Zofia Jeziorska (centre) and a female colleague (right), who were invited to the Zoo by a German couple (wife on the left) in around 1943. Both the Polish women had previously removed the "P" badge from their outer clothing that Polish forced labourers were compelled by law to wear.

The Berlin police had been growing concerned about the number of foreign visitors, and so in summer 1944 they asked the Zoo to issue a season pass so that an officer and an assistant could patrol the Zoo and monitor these "many foreigners".[106] Strangely, this request came from Department B3, the vice squad. It is possible that the police wanted to investigate potential "immoral" relations between Germans and foreigners visiting the Zoo together.

Forced labour at the Zoo

Forced labourers didn't only come to the Zoo as visitors, however. The conscription of a large proportion of its regular workforce had made it difficult for Zoo Berlin to remain operational. The management therefore actively sought to recruit forced labourers to replace the missing employees. At first, it was mainly Polish prisoners of war who were forced to work at the Zoo as "unskilled labourers". Later, they were joined by French POWs.[107] Every morning and evening, the French workers had to make their way to the Zoo by tram and by foot from their camp in Lichterfelde, around ten kilometres away. Initially they were overseen by two soldiers and later by a Zoo employee. By 1943, they were apparently registered as "civilian workers". Once the French workers

Two forced labourers from Poland on a visit to the Zoo

were no longer considered prisoners of war, their living conditions improved slightly. And yet they were still forced labourers: they could not refuse to work, could not change jobs, and certainly could not return home. The fact that they were apparently not mistreated at the Zoo does not alter the fact that they were there against their will. They were eventually housed in attic spaces around the equine zone, and did much of the clean-up after bombing raids.[108] During air raids, they may well have been allocated to various positions around the park with the task of putting out fires, even as the bombs rained down.[109] By mid-November 1943 at the latest, Heck submitted an application to General Building Inspector for the Reich Capital Albert Speer, requesting barracks be constructed to accommodate eighteen "civilian foreigners" – probably French forced labourers – on the site of the Waldschänke Inn, which had been bombed out in 1941. The barracks were to have a floor area of 61 square metres and be about two metres high.[110] The plan was approved, but it is questionable whether any real progress was ever made on the construction.

By late summer 1941, the labourers had been joined by six "young Ruthenian workers", whom Lutz Heck had "brought with him from the primeval forest of Bialowies".[111] It is not known exactly who those people were, though "Ruthenia" was once a name applied to Slavic lands. It is likely that Heck had these young people abducted during a visit to Göring's planned hunting grounds.

The Zoo archives contain a revealing document dating from January 1944 regarding payments to the tax office for November and December 1943. In addition to 8,177.07 Reichsmarks in payroll taxes, it appears the Zoo also paid 83.35 Reichsmarks as an "Eastern worker levy".[112] This levy was only paid for forced labourers from conquered areas of the Soviet Union. It can be assumed that these were the aforementioned "Ruthenians". Slavic workers were on the lowest level in the forced labour hierarchy when it came to food, accommodation and treatment.

In the meantime, the Zoo employed increasing numbers of forced labourers. After the bombing raids of November 1943, Albert Speer assigned 750 forced labourers (500 Soviet and 250 French prisoners of war) to perform clean-up work at the Zoo.[113] Later, the Zoo received a permanent contingent of "forty Badoglio Italians".[114] Pietro Badoglio was the new prime minister of Italy following the overthrow of the Fascist dictator Benito Mussolini, and he had arranged the ceasefire with the Western Allies. Italian soldiers captured by the Germans after that time were not described as "prisoners of war", as that would have granted them nominal protection under the Geneva Conventions. Instead, the Nazi regime referred to them as "Italian military internees" and deported hundreds of thousands of them to the Reich to work as forced labourers. In general, these workers were treated particularly badly as they were regarded as "traitors to the German-Italian axis". From September 1944, most of the Italians received "civilian worker" status. Although that meant minor improvements to their living conditions in the short term, it did not alter the coercive nature of the work, as was also the case for the French POWs.[115]

The use of forced labourers is also documented for other German zoos.[116] Not all zoos made explicit reference to such labourers in their publications, but it is likely that most used them to some extent. As the zoo in the Reich capital enjoyed such a high standing, it was probably assigned a relatively large number of forced labourers. And apparently the Zoo management happily accepted such offers – right to the bitter

February 1943
The German 6th Army is defeated at Stalingrad

end. As late as March 1945, the Zoo was still looking to recruit German war invalids to supervise a team of "twenty to forty Poles".[117]

The Zoo in flames

The 1943 financial year had started well for the Zoo, and it still had many visitors. Even large concerts were still taking place.[118] The Zoo did not experience a sharp decline in visitor numbers until the number of air-raid warnings increased and greater numbers of children were evacuated to the countryside.

Until then, the damage caused by the air raids had been manageable. But now, Lutz Heck was confronted with the need to make some important choices. Should he move the most important animals away and place them in other zoos out of reach of the Allied bombers? Or might that make his zoo unattractive to visitors? In 1943, the decision was taken to ship a large number of the animals out of Berlin. Most were housed in zoos that were considered less at risk, while some went to private individuals. A group of apes, for example, were accommodated in the cellar of zoologist Bernhard Grzimek's house in Johannisthal.[119] In total, around 500 animals of approximately 210 species were evacuated.[120] Among them was African elephant Tembo, who went to the supposedly safe Hellabrunn zoo in Munich in August 1943.[121] Unfortunately, she fell victim to an Allied bombing raid there. A report to the supervisory board stated that, in 1943, "about 150 large animals were housed outside of Berlin, primarily endangered species and valuable breeding groups."[122] However calm and self-assured that account may sound, the reality was very different. In mid-July 1943, Heck begged Viennese zoo director Otto Antonius to take his most important animals, as he feared "total loss". He offered Antonius several great apes, including the gorilla Pongo, four elephants, and "any rare birds he desired".[123] However, Antonius had nowhere to house these animals.

Pongo the gorilla, circa 1943

For Berliners, November 1943 was a month of pure terror. On the night of the 19th, they got a taste of the fearsome arsenal that the British RAF Bomber Command had been building up in the years since the German invasion of Poland. But the attack on 22 November eclipsed everything that had gone before. That night, more than

April 1943
Start of the Warsaw Ghetto Uprising

May 1943
Surrender of German troops in North Africa

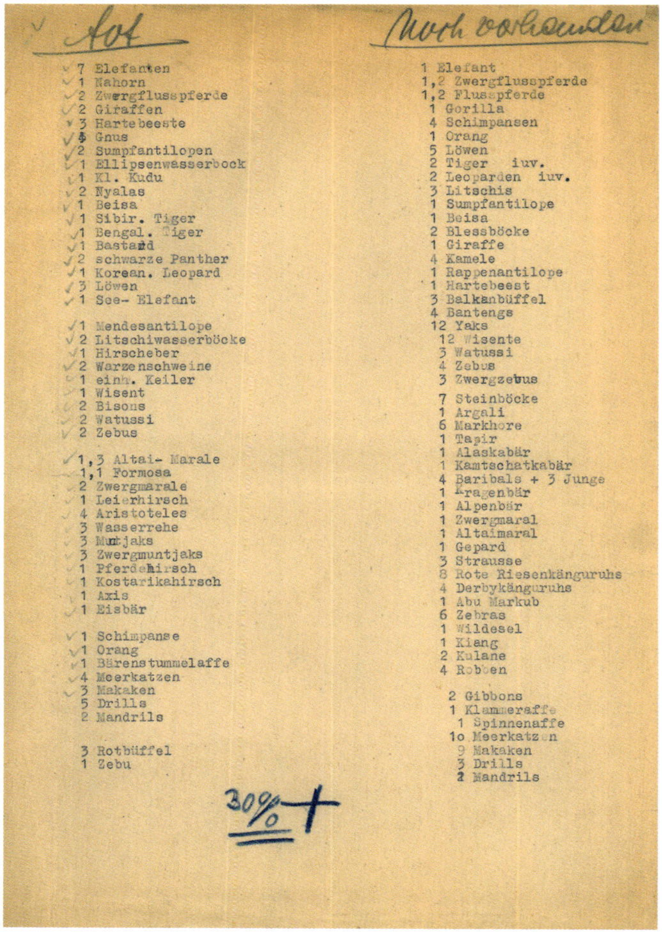

Tot

7 Elefanten
1 Nahorn
2 Zwergflusspferde
2 Giraffen
3 Hartebeeste
5 Gnus
2 Sumpfantilopen
1 Ellipsenwasserbock
1 Kl. Kudu
2 Nyalas
1 Beisa
1 Sibir. Tiger
1 Bengal. Tiger
1 Bastard
2 schwarze Panther
1 Korean. Leopard
3 Löwen
1 See- Elefant

1 Mendesantilope
2 Litschiwasserböcke
1 Hirscheber
2 Warzenschweine
1 einh. Keiler
1 Wisent
2 Bisons
2 Watussi
2 Zebus

1,3 Altai- Marale
1,1 Formosa
2 Zwergmarale
1 Leierhirsch
4 Aristoteles
3 Wasserrehe
3 Muntjaks
3 Zwergmuntjaks
1 Pferdehirsch
1 Kostarikahirsch
1 Axis
1 Eisbär

1 Schimpanse
1 Orang
1 Bärenstummelaffe
4 Meerkatzen
3 Makaken
5 Drills
2 Mandrils

3 Rotbüffel
1 Zebu

30% +

Noch vorhanden

1 Elefant
1,2 Zwergflusspferde
1,2 Flusspferde
1 Gorilla
4 Schimpansen
1 Orang
5 Löwen
2 Tiger iuv.
2 Leoparden iuv.
3 Litschis
1 Sumpfantilope
1 Beisa
2 Blessböcke
1 Giraffe
4 Kamele
1 Rappenantilope
1 Hartebeest
3 Balkanbüffel
4 Bantengs
12 Yaks
12 Wisente
3 Watussi
4 Zebus
3 Zwergzebus

7 Steinböcke
1 Argali
6 Markhore
1 Tapir
1 Alaskabär
1 Kamtschatkabär
4 Baribals + 3 Junge
1 Kragenbär
1 Alpenbär
1 Zwergmaral
1 Altaimaral
1 Gepard
3 Strausse
8 Rote Riesenkänguruhs
4 Derbykänguruhs
1 Abu Markub
6 Zebras
1 Wildesel
1 Kiang
2 Kulane
4 Robben

2 Gibbons
1 Klammeraffe
1 Spinnenaffe
1o Meerkatzen
9 Makaken
3 Drills
2 Mandrils

List of animals lost in the air raids, winter 1943/44

600 Halifax, Stirling and Lancaster bombers flew over the city, dropping in excess of 2,000 high explosives and incendiary bombs and devastating large parts of the western city centre. The Kaiser Wilhelm Memorial Church suffered extensive damage, and the adjacent Zoo became a sea of flames.[124] The bombs that night killed many of the Zoo's remaining animals. Among the victims were seven of the eight elephants, an African rhino, a chimpanzee, an orangutan, three lions, two tigers, two giraffes, and half of all the antelopes and deer.[125]

The Elephant Pagoda was reduced to rubble, the Antelope House burned down, and the buffalo and deer reserve with its wooden houses was consumed by flames. Other losses included the main restaurant, the administration building on Budapester Strasse, the director's residence, the animal hospital, and many of the workshops.[126]

However, no people were killed at the Zoo that night. The surviving animals were accommodated in makeshift animal shelters or taken elsewhere. The valuable shoebill, for instance, moved into the heated bathroom of a staff apartment next to the Pheasantry. This large bird survived until the end of the war, but sadly died shortly afterwards.[127]

The pygmy hippos were housed in the toilet block by the Elephant Gate, while sturdy boards were placed in front of the enclosure of the only surviving elephant and the cage of the male gorilla Pongo. The latter had actually escaped from his cage during the air raid and been found by ape keepers Mr and Mrs Liebetreu, who cornered him in the kitchen of their staff apartment. Thankfully, Pongo shuffled back into his enclosure of his own accord.[128] Given how strong the young ape already was, that episode could have ended badly.

No other dangerous animals escaped. The few animals found roaming the park were recovered by animal capture teams over the following days. Heck claimed that the reports of escaped animals were either "rumours or English propaganda". However, a wolf, a wapiti stag and several other injured animals had to be shot. The only animals to get away were smaller monkeys, birds and a pygmy hippo. The hippo was eventually found hiding in a corner of the flak tower.[129] In her later memoirs, veterinarian Maria von Maltzan, who also worked for Zoo Berlin and played an active role in the resistance against the Nazi regime,

July 1943
Mussolini is overthrown

Summer 1943
Evacuation of Berlin children to the countryside intensifies

reported how she once rescued a terrified little dingo after a night of bombing.[130]

The following night, Allied bombers again flew over Berlin, and the Zoo fell victim to the bombs once more. The Aquarium had already been damaged the night before, but now it was annihilated by an explosive that burst through the roof and detonated in the central Crocodile Hall. The glass of the tanks shattered and the tremendous volume of water washed reptiles and fish down the stairs and out onto the street. The animals that had not died immediately from the impact or the pressure now suffocated or froze to death in the cold November night. That night, Oskar and Katharina Heinroth were taking shelter in the cellar of the Aquarium, beneath the staircase by the entrance facing the Zoo. From the days of the Aquarium's construction, Oskar Heinroth had remembered this spot as particularly solid.[131]

Decades later, a Dutch former forced labourer named Lamers recounted his experiences of the night the Aquarium was destroyed and the days that followed. He had been deported to Berlin as a student in autumn 1943 and was forced to work in the Aquarium. On the night in question, he was in his accommodation close to the Zoo. He waited for the bombing to stop and then raced over to the Aquarium. The massive front door was lying on the pavement, water was pouring out onto the street, and an ankle-deep puddle had formed on the floor of the lobby. Lamers ran inside the building with his torch, hearing the sound of desperate animals flapping against the floor. At the foot of the grand staircase he found a pile of eight lifeless crocodiles. Only one crocodile, old Schwarzer Peter ("Black Peter"), had survived, thanks to his habit of resting under the concrete bridge that led through the Crocodile Hall. When daylight came, Oskar Heinroth and Inspector Seitz arrived on the scene with a rifle.

A shoebill takes shelter in a heated bathroom that survived the bombings, 1943

Shots could also be heard resounding from the Zoo, where other injured animals were obviously being put out of their misery. At the Aquarium, Schwarzer Peter, the Komodo dragon Moritz, a giant python, two giant tortoises, six gharials and some iguanas were the only surviving exotic animals, and they were rigid with cold. Heinroth and the staff tried to save them by placing them in the warm cellar until they could be taken to other zoos. Unfortunately, Moritz could not be saved, and the giant python also died after two days. Once they warmed up, the gharials became so animated and aggressive towards one another that only two survived the fracas. Lamers and a French prisoner of war named René were then forced to skin the python, and René made a tasty soup out of the crocodiles. The keepers were sceptical at first, but eventually they tucked in to the unusual dishes. The soup was also used to feed the students who, robbed of the carefree innocence of youth, had been employed as anti-aircraft auxiliaries at the flak tower on the northern edge of the Zoo.[132] In the end, Lamers was officially discharged by Heinroth and sent home.[133]

The Zoo, or at least its director, still had friends in high places. And so General Forestry Master Friedrich Alpers at the Reich Forestry Office readily provided loggers with chainsaws to clear away the toppled trees at Zoo Berlin, the commander of the nearby flak tower put together a team to extinguish the flaming piles of rubble, and a special veterinary troop spent days disposing of the elephant carcasses. A team from the Technical Emergency Corps blew up remaining parts of the Elephant Pagoda that were at risk of collapse and helped fix roofs. Reich Minister of Propaganda Joseph Goebbels, still also Gauleiter of Berlin, provided a contingent of soldiers.[134] As already mentioned, General Building Inspector for the Reich Capital Albert Speer provided 750 POWs as forced labourers, who were joined by "Italian military internees". A speedy clean-up was thus assured – thanks to the Zoo's outstanding significance for the Nazi propaganda effort and to Heck's own personal connections and high standing with the authorities. While the countless forced labourers were rewarded with nothing but poor living conditions, thirty Zoo employees received the "War Merit Cross second class with swords" for their "exceptional service combating damage caused by aircraft", and three others were rewarded with the "War Merit Cross second class without swords". Heck himself was

Dead elephants in the burned-out Elephant Pagoda, 1943

The ruined Aquarium, circa 1944

honoured with a "War Merit Cross first class with swords".[135]

In view of the constant air raids, Heck now asked the supervisory board to provisionally grant him sole management of the Zoo. This request blatantly contradicted the statutes of the shareholders' association and was rejected. Consequently, Heck contacted the Reich Forestry Office directly. Forestry Chief Göring promptly named Heck "Extraordinary Commissioner of the Zoological Garden" in a letter dated 16 December 1943.[136] And with that, Zoo Berlin had its very own "Führer".

Reopening under fire

The Zoo's first priority at this stage was to find provisional housing for all the surviving animals. This was largely provided by the animal houses that were still (partially) standing. In the damaged buildings, the few cages that were still intact were separated off with partitions from the destroyed parts, and stoves were installed for heating. Few buildings had not suffered damage, and all the feed reserves were gone. The animals left alive needed to be fed, and the Zoo's hundredth anniversary was fast approaching. Could the centenary of Germany's oldest Zoo be allowed to pass by unobserved? After all, Lutz Heck was the most important and most powerful zoologist in

January 1944
Allied troops land in Anzio, south of Rome

Spring 1944
Heavy Allied bombing raids on Berlin

Shard from one of the glass panes in the Crocodile Hall

Following an air raid, a surviving tapir warms itself on a smouldering pile of debris

the Third Reich, and the Zoo had been classified as essential to the war effort given its function of improving morale among the people of Berlin. What would become of the animals and the keepers if the Zoo remained closed? And so a reopening was planned, in the middle of the most devastating war the world had ever seen, with unremitting air raids. Over the coming months the Zoo grounds were bombarded repeatedly. One attack in February 1944 was particularly devastating. More animals were killed – including public favourites like elephant seal Roland III, who passed away on 1 March 1944 from wounds sustained from an exploded shell.[137] Nevertheless, a considerable animal collection remained. At the end of 1944, the Zoo still had almost 1,700 animals of more than 500 species.[138]

Berlin's zoological garden reopened on 25 July 1944 – after Allied troops had landed in Normandy, and with the Red Army poised to occupy East Prussia. Heck agreed security measures with the chief of police: in the event of an air-raid warning, the closest police station would send officers to evacuate the Zoo grounds. Also, the commander of the flak tower agreed that Zoo visitors would be able to take shelter there. To ensure there would be enough space, the maximum number of visitors each day was set at 5,000. A wide evacuation path to the overground bunker was created next to the Pheasantry. Heck even endeavoured to borrow interesting animals from elsewhere to improve the Zoo's appeal.[139]

In the end, Zoo Berlin's hundredth anniversary was celebrated with very little pomp. The small number of invitees were treated to a film screening and a tour. Afterwards, six employees were awarded with the "War Merit Cross second class with swords" by a representative of the city government. The occasion was rounded off with a dinner and social evening. The guests of honour included the prime minister of the state of Braunschweig, General Forestry Master and *SS-Obergruppenführer* Dietrich Klagges, some colleagues from the Reich Forestry Office, including Master

June 1944
Normandy landings

20 July 1944
Wehrmacht officers attempt to assassinate Hitler

of the Hunt and *SS-Brigadeführer* Ulrich Scherping, the directors of the zoos in Leipzig, Cologne and Hamburg, and some high-ranking local personalities and academics. Reich Minister of Propaganda Joseph Goebbels composed a long telegram, and Göring sent his regards.[140] This shows that Director Heck was on good terms with the regime's top brass until the very end. In January 1945, Heck was a guest at Göring's birthday dinner in his country residence Carinhall on the Schorfheide hunting estate.[141]

In their reporting on the reopened Zoo, the newspapers had a few vague words of praise for the reconstruction efforts, but their primary focus was what animals could still be seen. Berliners' interest in the Zoo seemed unabated. The respite from Allied bombing raids granted in the second half of 1944 brought the heavily damaged Zoo around 250,000 visitors. On Sunday, 10 September 1944, just under 12,000 people made their way to the Zoo. One of the main attractions was Pongo the gorilla.[142]

On the front line

The Second World War had been set in motion by megalomaniacs in Berlin, and now it came full circle as the war in Europe ended right back where it started. In the night of 20 to 21 April

The Lion Gate on reopening day in August 1944

27 January 1945
The Red Army liberates Auschwitz

February 1945
The Allies hold the Yalta Conference

7 March 1945
Allied forces take Cologne

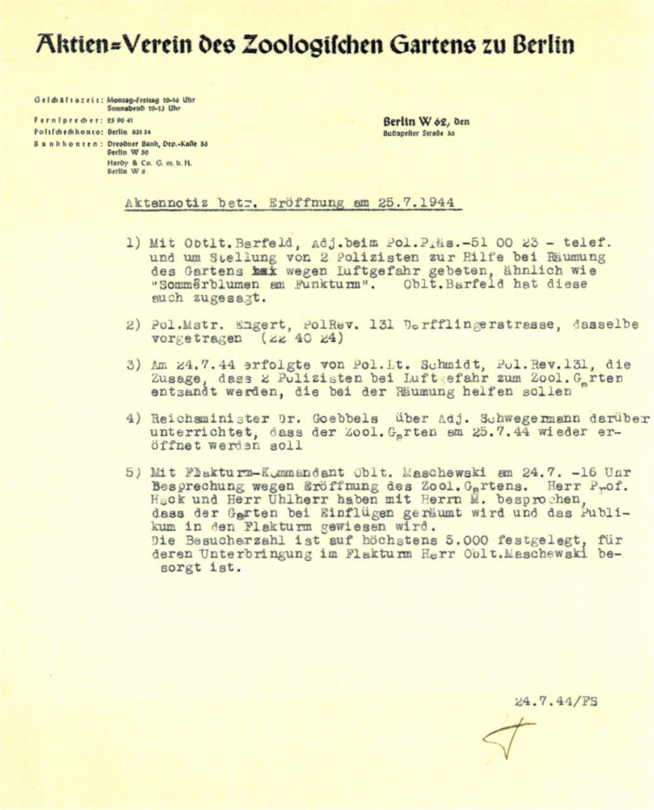

Aktien=Verein des Zoologischen Gartens zu Berlin

Geschäftszeit: Montag-Freitag 10-16 Uhr
Sonnabend 10-13 Uhr
Fernsprecher: 25 90 41
Postscheckkonto: Berlin 831 34
Bankkonten: Dresdner Bank, Dep.-Kasse 36
Berlin W 50
Hardy & Co. G. m. b. H.
Berlin W 8

Berlin W 62, den
Budapester Straße 36

Aktennotiz betr. Eröffnung am 25.7.1944

1) Mit Oblt.Barfeld, Adj.beim Pol.Präs.-51 00 23 - telef. und um Stellung von 2 Polizisten zur Hilfe bei Räumung des Gartens bei wegen Luftgefahr gebeten, ähnlich wie "Sommerblumen am Funkturm". Oblt.Barfeld hat diese auch zugesagt.

2) Pol.Mstr. Engert, PolRev. 131 Derfflingerstrasse, dasselbe vorgetragen (22 40 24)

3) Am 24.7.44 erfolgte von Pol.Lt. Schmidt, Pol.Rev.131, die Zusage, dass 2 Polizisten bei Luftgefahr zum Zool.Garten entsandt werden, die bei der Räumung helfen sollen

4) Reichsminister Dr. Goebbels über Adj. Schwegermann darüber unterrichtet, dass der Zool.Garten am 25.7.44 wieder eröffnet werden soll

5) Mit Flakturm-Kommandant Oblt. Maschewski am 24.7. -16 Uhr Besprechung wegen Eröffnung des Zool.Gartens. Herr Prof. Heck und Herr Uhlherr haben mit Herrn M. besprochen, dass der Garten bei Einflügen geräumt wird und das Publikum in den Flakturm gewiesen wird.
Die Besucherzahl ist auf höchstens 5.000 festgelegt, für deren Unterbringung im Flakturm Herr Oblt.Maschewski besorgt ist.

24.7.44/FS

Memo regarding agreements made with the police and the commander of the flak tower at the Zoo, 24 July 1944

1945, the Royal Air Force dropped bombs on Berlin for the last time. But what came next was even more calamitous. Over the weeks that followed, the Red Army tenaciously conquered Berlin street by street, suffering many casualties in the process. Adolf Hitler had given the lunatic order to his *Volkssturm* militia to "fight to the last man", insisting on a senseless battle of annihilation in a war that had long since been lost rather than allowing the German people to "suffer the indignity" of surrender. Now, the flak tower and the zoological garden next to it became the targets of Soviet ground troops. The Zoo finally closed its gates on 22 April 1945. The remaining Zoo employees were drafted into the *Volkssturm* militia of old men and teenagers and found themselves forced to dig trenches and erect anti-tank barriers across the gardens that had once been their workplace.[143] Lutz Heck had previously hired his son, also named Lutz, as a junior assistant, probably to protect him from being drafted.[144] The Zoo director later claimed to have saved his staff by using all the means at his disposal to ensure that they were deployed under his command to defend the Zoo.[145] This claim could not be confirmed.

The influential zookeeper had so far been able to prevent the killing of predators at his zoo. In the final days, however, Heck finally ordered the shooting of the four remaining adult lions. The two cubs were taken to the unused air-raid shelter under the Ibex Rock. Since the enclosure of the last remaining brown bear had been damaged, it too had to be dispatched. Heck himself apparently shot dead an escaped baboon that he had originally brought back with him from an expedition in Cameroon.[146]

With its large anti-aircraft artillery, the flak tower at the Zoo now directed its fire at Soviet tanks and artillery in city districts some distance away. And yet the unstoppable Red Army fought its way through the government quarter and the Tiergarten park right up to the gates of the Zoo. It probably entered the grounds on 30 April. Several thousand Wehrmacht soldiers were positioned in the Zoo's buildings and environs.[147] First, the Soviet soldiers of the 39th and 88th Guards Rifle Divisions succeeded in conquering the Aquarium and the large restaurant, where Wehrmacht units were ensconced. Advance troops then managed to blast their way through the barricades and the fence, allowing the Soviets to gain fire support from tanks and ordnance and finally attack the massive bunker and flak tower. Soviet tanks were soon stationed by the Hippo House, shelling the

19 March 1945
Hitler orders the destruction of German infrastructure

16 April 1945
The Red Army breaks through the German defences on the River Oder

4 May 1945
The Wehrmacht surrenders in northern Germany and Scandinavia

nearby tower from close range.[148] On 2 May, the tower's crew surrendered.

For several days, the Zoo was transformed into a battlefield. The few keepers who had apartments on the Zoo grounds remained in permanent hiding in the small underground bunker at the Elephant Gate to escape the artillery fire. During pauses in the fighting, usually at night, they would risk their lives by emerging from the shelter to care for the animals. On one such sortie, ungulate keeper Hubert Eberhard was killed by shrapnel.[149]

Human and animal corpses littered the grounds.[150] The Zoo's director, "Professor" Heck, left the Zoo on 30 April, shortly before the Russian troops arrived, along with all delegates and assistants.[151] Heck had obviously prepared his escape in advance, as well as a future life for himself far from the Reich's fallen capital. In early March, he had placed a special order for eight metal cases to store "important archives of the Zoological Garden and the Reich Forestry Office". Perhaps they were used to transport the files to Schorfheide.[152] Heck also removed large parts of his private library and his works of art. Ultimately, Heck was no different from all the other "golden pheasants" – a pejorative nickname for high-ranking Nazis who had been awarded the Golden Party Badge for outstanding service, but who went into hiding while the ordinary people around them were dying.

After the war, Heck gave various accounts of how and when he left the Zoo and Berlin. On one occasion, he claimed that on 30 April 1945 he and the Zoo's *Volkssturm* fighters had been ordered into the flak tower, which he did not leave until 2 May.[153] In another version of the story from 1951, he stated that he did not depart from Berlin until 7 May, when the Zoo's driver was named its new manager. Heck claimed that he had no choice but to escape as his involvement in the *Volkssturm* would constitute grounds for execution.[154] This last claim was no more accurate than the alleged immediate appointment of a new Zoo manager. It is much more likely that Heck came up with these justifications for his actions retrospectively. The entire Zoo management board knew that it would be considered culpable. Heck's position within the National Socialist apparatus, his relationship with Hermann Göring and his Party membership all meant that he could not expect to have any future at Zoo Berlin. He must have felt that his chances for a fresh start were better in what was likely to become the American zone of occupation.

Heck may also have feared that the loot he had secured during the Nazis' eastern expansion might now backfire on him.[155] Keeper Elisabeth Johst certainly remembered Soviet officers asking her father, the zookeeper Heinrich Schwarz, at gunpoint about the Zoo managers, whose names were noted on a "blacklist".[156]

The memoirs of Katharina Heinroth, who was to be Heck's post-war successor as the director of Zoo Berlin, record the moment of his departure as follows: "When the Russians were advancing on Berlin, Lutz Heck bade me farewell, telling me that he had to leave, and asked me to look after the wives and children of the Zoo personnel. Shortly afterwards, his delegate Pfeiffer stormed in, saying that he too had to 'make tracks'. I attached a cloth with a large red cross painted on it to the outside of the bunker door in hopes of keeping the hostilities at bay."[157] One day later, Soviet soldiers reached the bunker on Budapester Strasse. On 2 May 1945, the battle for Zoo Berlin was over.

7 May 1945
Germany surrenders to the Western Allies

8/9 May 1945
Germany surrenders to the Red Army

Emerging from the ruins, 1945–1949

An imperfect peace

Finally, on 3 May 1945, all German troops in Berlin surrendered, leaving the city and the Zoo one vast field of rubble. All the animal houses and other buildings had been destroyed or severely damaged, and the zoological garden that had once housed more species than any other in the world now had just a handful of surviving animals. Also, many employees were now prisoners of war, dead, or missing.

At the Hippo House, a dead hippopotamus floated in the water, an unexploded rocket protruding from her body; next to the corpse, a smaller hippo was trying to cool off in the remaining water.[1] That hippo was the young bull Knautschke. Miraculously, three hippos had survived the destruction of the Hippo House during the bombing on 22 November 1943.[2] But now, only Knautschke was left, paddling around the body of the larger hippo. His delicate skin was bruised, parched and crusted over. The explosions had caused the pool to leak, and the water was draining away. In the following weeks, keepers had to haul buckets of water over to the pool from the few remaining pumps that still worked in order to cool the young hippo down and keep his skin damp. But the most important thing was: he was alive! The remarkable story of Knautschke, the hippo who survived the war, would stay with the Berlin public for many years to come. Soon people started dropping by the Zoo to share their meagre food rations with the celebrity hippo and the other surviving animals.

On that day in May, Red Army soldiers stood around the Monkey Rock watching the few surviving baboons. Meanwhile, in the neighbouring Ape Conservatory, a gorilla and a chimpanzee lay dead alongside three dead SS officers. The gorilla was Pongo, successor to the celebrity ape Bobby. Pongo had two stab wounds to his chest.[3] Had the SS men perhaps killed the primates with a bayonet in order to obtain a safe hiding place for themselves, only to be struck down by gunfire from Soviet soldiers?

Removal of the many human and animal corpses was the immediate priority, but completing that task took almost two weeks. Employees noted down the burial sites of two civilians, fifty-eight Wehrmacht soldiers, and six members of the Red Army.[4] Most of those bodies were exhumed over the next few years and reburied in

Kathi the hippo at Zoo Berlin, 2018

Hippopotamus

Despite the fact that its name is Greek for "river horse", the hippopotamus (*Hippopotamus amphibius*) does not actually share a lineage with horses, but with pigs and whales. At up to four and a half metres long and over three tonnes in weight, these huge herbivores have to consume a lot of food. Their diet in the wild consists mainly of short grass and some aquatic plants. At Zoo Berlin they consume up to 45 kilos of green fodder daily, in addition to turnips and concentrated feed. During the commentated feeding sessions, keepers also sometimes give them bread as a treat. However, one should think twice before calling the hungry, hungry hippos greedy, as they actually eat only around 1.5 per cent of their body weight per day, which is half that of similarly sized mammals like rhinoceroses.

Hippos Kathi and Max, 2016

Hippo Kathi with her calf Max, 2014

Hippos spend their time in two different types of habitat: during the day they keep to the water, but at night they go ashore and sometimes wander miles inland to find the best grazing spots. One reason why they submerge themselves in water during the day is to protect their sensitive skin. Hippos also secrete a reddish sweat-like liquid that acts as a sunscreen for exposed areas. The reddish-brown colour this liquid takes on when it dries is the basis for the old myth that hippos sweat blood.

Some male hippos establish territories and defend their sole right to mate in those areas against younger challengers. Their massive, sharp canine teeth can easily cause fatal injuries. Hippos also aggressively protect their offspring, which makes them dangerous to potential predators and humans.

Female hippos often live in groups, but these are non-hierarchical and prone to change. An apt description of hippos is "sociable loners", which is one of several reasons why they are usually rather uncomplicated zoo inhabitants. Hippos rarely fall ill and are quite undemanding. They can even reach a grand old age of around fifty in human care. Today, almost all Berlin's hippos are descendants of the legendary hippopotamus bull Knautschke, one of only around ninety animals at the Zoo that survived the Second World War. Knautschke fathered thirty-two offspring in his lifetime before eventually dying in 1988 from injuries sustained during a fight with his son Nante.

The modern Hippo House is a visitor highlight at Zoo Berlin – it is even used as a wedding venue. A special feature of the habitat is the long glass pane that enables visitors to watch the enormous mammals under the water. Technically, hippos don't actually swim; instead, they run underwater, pushing themselves off the bottom with their short legs. Because they weigh much less in the water, "swimming" hippos look positively graceful.

cemeteries. At the Elephant Gate a mass grave was dug for twenty-five German soldiers. The remains, buried only about 50 centimetres deep, gave off such a foul stench that even once they were exhumed the soil there had to be replaced.[5]

The day the guns of World War II fell silent by no means marked an end to the violence in and around the Zoo. The occupying forces now had to consolidate their power and establish a provisional administration among the rubble. They were plagued by fears of a guerrilla war waged by remaining Nazi *Werwolf* troops.

What's more, German civilians were now confronted with the concentrated anger of the Soviet victors. The Germans had invaded their country, murdered millions of civilians, deported them to work as forced labourers, and starved them to death. Wehrmacht soldiers and SS officers had slaughtered millions of Soviet prisoners of war or left them to rot in jail. Now, the Soviets got their revenge: they raped thousands of women, plundered the city, and shot anyone who put up resistance. Assuming the Zoo keepers were members of the military because of their uniforms, Red Army soldiers arrested them, dragging them from the bunkers and ruined buildings. Civilians were robbed of all their valuables.[6]

After the first chaotic weeks, in many parts of Berlin the situation improved a little. Although they had been bombed out themselves, the few remaining Zoo members of staff self-sacrificingly took care of the animals, initially without any pay but with the permission of the Russian commanding officer.

Oskar Heinroth died of pneumonia on 31 May 1945. His place of death was his beloved but now destroyed Aquarium. His wife Katharina had taken him back there after the Soviet soldiers had withdrawn. The Red Army had occupied the severely damaged building over a period of three weeks, leading to further destruction and looting. The Zoo's carpenter carved a coffin for Heinroth out of damaged wooden doors. Weeks later, newly appointed Aquarium director Werner Schröder picked up Heinroth's ashes in a cardboard box from the crematorium.[7] The ashes were buried at the Zoo on 15 August 1945.[8] Since 1949, a plaque on the Zoo side of the Aquarium has commemorated the great biologist. In 1989, his wife Katharina Heinroth was laid to rest next to him.

During the Nazi period, the Heinroths had remained apolitical or sceptical. They were not members of the NSDAP but neither were they involved in the resistance. They did, however, closely follow the proceedings of the People's Court against their friend, zoologist Walther Arndt, for making critical remarks about the Nazi regime, and participated in the request for clemency. The People's Court had been set up by the Nazis specifically to pass high-profile judgements against opponents.

Taking stock

In May 1945, an unknown Zoo employee compiled an inventory of the few surviving animals. The list was prepared in German and Russian for the relevant commanding officer of the occupying forces and shows the full extent of the losses. On 31 May 1945 the Zoo's entire stock totalled ninety-two animals, including a workhorse.[9] Katharina Heinroth later recalled that there were eighty-one surviving animals.[10] The numbers were subject to fluctuations, as in the chaos injured or sick animals died, and other animals were stolen. The fences and border walls were, after all, full of gaping holes. The only big cats that had survived the war were the two young lions Sultan and

13 May 1945
First performance by the
Berlin Chamber Orchestra

21 May 1945
The *Berliner Zeitung* newspaper
is published, under Soviet control

26 May 1945
First performance by
the Berliner Philharmoniker

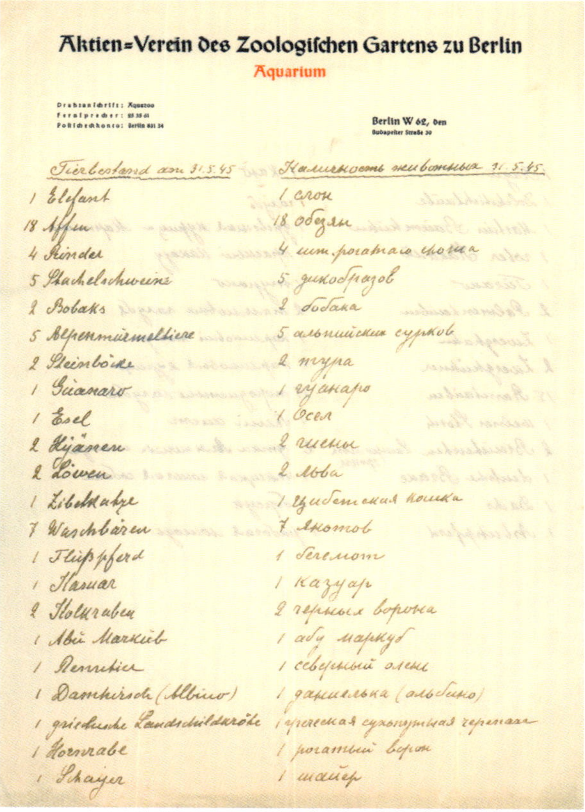

Aktien=Verein des Zoologischen Gartens zu Berlin

Aquarium

Drahtanschrift: Aquazoo
Fernsprecher: 25 35 61
Postscheckkonto: Berlin 831 34

Berlin W 62, den
Budapester Straße 30

Tierbestand am 31.5.45	Наличность животных 31.5.45.
1 Elefant	1 слон
18 Affen	18 обезьян
4 Rinder	4 шт. рогатого скота
5 Stachelschweine	5 дикобразов
2 Bobaks	2 бобака
5 Alpenmurmeltiere	5 альпийских сурков
2 Steinböcke	2 тура
1 Guanaco	1 гуанако
1 Esel	1 Осел
2 Hyänen	2 гиены
2 Löwen	2 льва
1 Zibetkatze	1 цибетская кошка
7 Waschbären	7 енотов
1 Flußpferd	1 бегемот
1 Kasuar	1 казуар
2 Kolkraben	2 черных ворона
1 Abu Markub	1 абу маркуб
1 Renntier	1 северный олень
1 Damhirsch (Albino)	1 даниэлька (альбино)
1 griechische Landschildkröte	1 греческая сухопутная черепаха
1 Hornrabe	1 рогатый ворон
1 Schajer	1 шайер

List of surviving animals in German and Russian, 1945

Bussy. They were fortunate enough to have been transferred from the larger Predator House to the smaller one as tiny cubs in 1943. While the larger facility had been almost completely destroyed in the devastating air raid of November 1943, its smaller counterpart had remained largely undamaged.[11] Then, when the fighting approached the Zoo in 1945, the keepers hid the lions in the bunker space at the Ibex Rock.[12] That meant they escaped the attention of the marauding Russian soldiers, who had shot some animals and taken others away as mascots. Several animals, including a bear, were slaughtered and eaten by Red Army soldiers in the days following the city's liberation from the Nazis.[13] During May 1945, the commandant's office confiscated several baboons, a turaco and some gamebirds.[14] On 28 May 1945, when Berlin had been liberated not quite a month and was still entirely in the hands of the Red Army, a representative of the Zoo asked the Soviet city commandant General Nikolai Berzarin to stop his troops from hunting the few surviving Zoo animals.[15]

The remaining animals were all housed in the still extant Poultry House, and the staff were tasked with performing the most urgent clean-up work. First of all they had to collect weapons and ammunition and bury them in shell craters, under the supervision and constant threats of Soviet soldiers. In the future, these remnants of war would emerge again and again during excavation work. This was also true of the remains of Wehrmacht soldiers who were buried with their weapons.[16] And of course it was not only the animal houses and the fencing that were ruined, all the underground pipes and the sewer system had also been destroyed, and the water tower was peppered with no fewer than sixty-eight holes.[17]

Funds did still exist, but the Zoo did not have access to them. Before fleeing, Zoo director Lutz Heck had paid himself and some of his employees generous advances and transferred half a million Reichsmarks to bank accounts in Bavaria and Baden-Württemberg. Many people had already realised that those states would not be occupied by the Red Army, but by US troops. Heck later insisted that he had done this in order to keep the Zoo's funds safe. However, at the time in question he was not interested in making that money available to the Zoo. In November 1945, refusing to accept his dismissal, he described himself as director of the "southern branch" of the Zoo's shareholder association, which was now "in exile".[18]

A new leader was needed. The first few days after the end of the war involved a good deal of improvisation. A Soviet officer gave bird keeper

June 1945
Re-establishment of the Social Democratic Party of Germany (SPD)

June 1945
Founding of the Christian Democratic Union of Germany (CDU)

11 June 1945
Re-establishment of the Communist Party of Germany

Heinrich Schwarz the task of procuring feed.[19] The deputy mayor of the Tiergarten district then appointed Schwarz and a garden inspector named Karkowsky as acting directors of the Zoo. However, since both of them had probably been members of the NSDAP, they had to be replaced. Next, the mayor of Schöneberg and later of the Tiergarten district appointed Ernst Übrick, the last driver of the Zoo, as acting director. Meanwhile, a man named Edmund Schirrmacher, who had been a waiter at the main restaurant, applied to the Soviet commandant's office for the job of director. He was appointed the head of clean-up operations by Berlin's deputy mayor Paul Schwenk on the orders of Soviet General Berzarin. Ultimately, Übrick and Schirrmacher went on to manage the Zoo's operations as acting members of the works council and to make joint decisions about what to do with the remaining cash.[20] The two men initially shared the tasks; Schirrmacher procured the feed and all expenses were authorised jointly. Übrick oversaw clean-up work performed by former NSDAP members who were enlisted for the task by the Schöneberg district office.[21] Later, he also organised the first *Trümmerfrauen* ("rubble women") assigned to the Zoo by the authorities.

Katharina Heinroth, who was actually eminently qualified to become the next director herself, was motivated by her strong sense of duty towards the Zoo to provide help and advice – despite the fact that Übrick and Schirrmacher were apparently constantly at odds and often drunk.[22] To make sense of this chaos, we must try to picture what things were like in Berlin during those months following the liberation: telephones did not work, roads were still impassable in many places, and public transport and mail services were very slow to resume. And yet, despite all those difficulties, Berlin's zoo reopened its gates on 1 July 1945, a little more than a month after the war ended.

Visitors sun themselves on a visit to the severely damaged Zoo, summer 1945.

Just as miraculously, the tiny remaining animal stock began to grow again. In mid-August the Zoo took on a lion and an Asian black bear from a circus that had been disbanded by the Soviet occupying forces. It also received wolves, foxes, raccoons and domestic animals.[23] Of course, once it had these new animals, the Zoo also needed to acquire more feed. For one month in late summer 1945, Heinroth calculated a requirement of 3,000 kilograms of hay, 1,000 kilograms of straw, 3,500 kilograms of fodder beets, 550 kilograms of offal, 650 kilograms of horse meat, and 600 kilograms of potatoes. She ap-

11 July 1945
Beginning of four-power rule in Berlin

17 July 1945
First day of the Potsdam Conference

August 1945
The US drops two atomic bombs on Japan

Surviving lions on their "steppe" with bombed-out buildings in the background, circa 1946

proached the Allied Kommandatura, which in November 1945 instructed the city's mayor to ensure that Zoo Berlin receive sufficient feed for its animals, given the cultural significance of the Zoo and the need to maintain it. However, the instructions also stipulated that, as far as possible, the feed should be unfit for human consumption, and the Zoo was not allowed to take in any more animals without first obtaining permission from the Kommandatura.[24] In general during those years, one gets the impression that both the Soviet and later the British occupying forces were sympathetic towards the Zoo. Both repeatedly helped out by granting special approvals.[25]

Katharina the Great

In late July 1945, a representative of the mayor of Berlin who had inspected the ruined Zoo made his recommendations. He meticulously listed the damage and necessary measures for each building.[26] It was clear that the majority of the buildings could not be rebuilt. Nevertheless, the city hall saw potential for the Zoo if it were able to re-establish itself as a scientific facility.[27]

And that is what happened on 3 August 1945, with the appointment of highly educated zoologists Katharina Heinroth as zoological director and Werner Schröder as business director. In the

2 September 1945
Japan officially surrenders; end of World War II

27 September 1945
Publication of the first edition of the *Tagesspiegel* newspaper

18 October 1945
Start of the Nuremberg trials against leading Nazi war criminals

past, Heinroth had frequently contributed to her husband Oskar's research and she felt a strong emotional connection to Zoo Berlin. Therefore, she did not hesitate to say yes when she was asked to take over and rebuild the Zoo. With Heinroth's agreement, Schröder was tasked with assisting her as business director. The dynamic duo went on to achieve great things for Zoo Berlin. Although they were both shy, retiring types, they were remarkably strong willed and complemented one another very well. On 26 September 1945, the newly constituted – though not yet fully legitimised – supervisory board confirmed Heinroth and Schröder in their offices.[28] Katharina Heinroth became Germany's first female zoo director.

Like most zoo directors, the petite Heinroth made her rounds of all the Zoo's departments every morning. On one of those rounds, she adopted the huge Newfoundland dog Ajax, who

Katharina Heinroth doing her morning rounds with the Newfoundland dog Ajax, circa 1946

10 January 1946
First session of the United Nations General Assembly

20 January 1946
Professors are reinstated at Berlin's Friedrich-Wilhelms-Universität (soon to become the Humboldt-Universität zu Berlin)

Katharina Heinroth with a pig-tailed macaque on her return from an expedition to Borneo, 1955

had been handed over to the Zoo in the difficult days immediately following the war. Her predecessor Lutz Heck had always done his rounds with a stately bearing and a deferent entourage following several paces behind – a habit that had earned him the nickname "the Prince". Now, Heinroth was fondly but respectfully nicknamed "Katharina the Great" by her employees.[29] Later, the Tierpark's founding director Heinrich Dathe dubbed her "Katharina the inimitable".

For certain other former Zoo employees, reinstatement was unthinkable – at least for the time being. In March 1946, at the authorities' request, Heinroth reported former director Lutz Heck, scientific assistant Hermann Steinmetz, and other staff members as no longer employable due to their membership in the NSDAP.[30]

There was still a severe lack of animal feed, so all options had to be exploited. One Zoo employee drove regularly to the Schorfheide hunting estate in the summer and autumn of 1945 to mow grass for the Zoo. He got caught, and Heinroth had to explain herself to the local forestry administration. The Zoo management had probably assumed that Hermann Göring's former private grounds were no longer under supervision now they had no owner.[31] In the winter of 1945/46, Berliners brought a large number of pet budgerigars, European pond turtles and North American terrapins to the Zoo, probably because they were no longer able to keep them at home due to a lack of heating and feed. The Zoo had already publicly requested such "gifts" in June 1945.[32] Another mass donation of pets occurred in autumn 1946. The Zoo was able to sell some of those animals on to wealthier Berliners.[33] At the end of the first year of peacetime, Heinroth was able to exhibit 222 individual animals of around ninety-five different species. These were mostly single specimens, with very few breeding pairs. But there were a few exceptions: twenty-five Hermann's tortoises, twenty domestic pigeons, ten raccoons, and ten surviving baboons from the once-so-spectacular troop brought back by Lutz Heck from one of his expeditions.[34]

Before any animal houses could be rebuilt, first the rubble had to be cleared from the grounds. By the time the Zoo reopened, a few makeshift pathways had been cleared. Thousands of cubic metres of rubble lay piled up in the enclosures and scattered across the paths and parkland. About 200 of the famous "rubble women" worked to shovel the debris into open wagons on a specially built temporary railway.[35]

7 March 1946
Founding of the Free German Youth movement (FDJ)

22 April 1946
Socialist Unity Party of Germany (SED) created from forced merger of the KPD and SPD in the Soviet zone

Like everywhere else in Berlin, this clean-up work was very dangerous, and the inexperienced women and men almost always worked without adequate protective equipment. Several fatal accidents occurred at the Zoo during this time due to falling beams or live munition concealed in the rubble.[36] In summer 1946, the municipal government stopped the support payments for rubble clearance. Heinroth was able to continue the task with a group of workers consisting of ten women and two men thanks to a generous discount offered by a local businessman.[37]

Heinroth did for her zoological garden what practically all Berliners were doing at that time: she got things organised, she made things happen. The materials needed to keep the Zoo operational had to come from somewhere, and often it was in the form of donations. For instance, Heinroth was happy to issue two free tickets to the daughters of a man who donated roofing material, which was urgently needed, for example to cover an anteroom at the Hippo House.[38] In the first half of the year after the reopening, more than 300,000 visitors came to the Zoo despite its ruinous condition and the bad transport connections.[39]

In the early post-war years, the citizens of Berlin were constantly hungry and always on the lookout for ways to obtain food. Many people planted vegetable gardens. Immediately after the war, the people of Berlin had turned the Tiergarten park into a crop field. The trees had been felled for firewood, so the area was clear for planting potatoes, turnips and other vegetables, which helped Berliners survive the winter. The British occupying forces also wanted all the trees in the Zoo to be chopped down so the area could be used exclusively for growing vegetables and keeping farm animals. However, Heinroth fought bravely for her trees and exotic animals.

"Rubble women" clearing up at the Zoo, circa 1945

She also tried to play off different offices and authorities against one another.[40] In general, Heinroth prevailed, and some of those old trees can still be seen at the Zoo today. The exceptionally harsh winter of 1946/47 went down in history as the "hunger winter", and 1947 was to be the most austere year of the post-war period. Many Berliners took foraging trips known as "hamster trips" out to local farms, and the black market flourished.

Heinroth permitted Zoo employees to use free space to grow vegetables for their own consumption as well as fodder plants for the animals. She and Schröder personally helped out with growing and harvesting the crops, and Heinroth tried to

5 September 1946
US sector radio station RIAS
begins broadcasting

20 October 1946
Last joint city council elections
in Berlin

get land for fodder cultivation in the surrounding area.[41] She frequently had to take action against visitors and employees who were caught gathering firewood at the Zoo without authorisation. The Zoo needed that firewood itself, to heat the animal accommodations. The Zoo fences still had holes, and the records show multiple cases of burglary reported to the police. In late 1946, Heinroth reported the disappearance of several mallards and domestic ducks, and then a woman was caught stealing vegetables.[42] The following year, duck thieves were caught red-handed.[43] Children and other unauthorised individuals came and went unchecked.[44] Thieves also broke into the cash offices, and in February 1947 a deer was killed. The Zoo offered a reward for information leading to the apprehension of the culprits.[45] In May 1947, a particularly audacious gang scored a coup when they arrived at the administration and utilities area with a horse and cart and claimed that they had something to pick up. While the gatekeeper went to inquire, they escaped with 150 garden chairs.[46] Because of the repeated thefts, for a short time the local police station provided the Zoo with a two-man patrol team. Soon after their departure, however, the thefts and animal killing began again.[47] This suggests that people who worked at the Zoo were involved. Those suspicions were reinforced when elephant meat turned up on the Berlin black market shortly after the death of elephant Siam. The elephant's flesh had actually been intended as feed for some of the remaining predators.[48] Bombed-out employees were housed in many places around the Zoo – even the Ibex Rock bunker contained a makeshift apartment.[49]

Heinroth was aware that she had to procure new animals as soon as possible in order to make the Zoo more appealing. The purchase of exotic animals was, of course, out of the question. So in spring 1946, she submitted a request to the administrative director of Berlin's city forests. In order to at least be able to "exhibit the native animal kingdom as comprehensively as possible" she asked for swans, ducks and other waterfowl, birds of prey, corvids, a roebuck, wild boars, foxes, martens, polecats, weasels, young hares and wild rabbits.[50] And she did indeed receive some of those animals.

The Zoo also had some gifts that seem rather odd from our modern perspective. At that time, the Cold War was still lukewarm and the city was not yet divided. The State Party School of the Socialist Unity Party of Germany (SED) donated a young fox to the "Zoo of the German capital". Heinroth was only moderately enthusiastic about this, as she had been offered many foxes.[51]

There was, however, another problem that she had been dealing with since the early months of her directorship. As mentioned above, Lutz Heck had transferred generous advances or pension payments to his closest employees, his father and himself from the Zoo's coffers to an account in southern Germany. There, in "exile", he continued to refer to himself as the director of Zoo Berlin. Heinroth learned from contacts in Stuttgart that Heck was planning to open a "branch" of the Zoo down south with the money he had squirrelled away.[52] She therefore had to make sure that all German zoologists knew that Heck was in fact no longer director of Zoo Berlin. She also had to fight for years to get the money back.

Two steps forward, one step back

Both Werner Schröder and Katharina Heinroth worked tirelessly for their zoo, but a letter written by the latter to Munich zoo director Heinz Heck (Lutz Heck's brother) in August 1947 shows

The demolished flak tower, 1949

the frustration she often felt. The topic of the letter was the search for a successor to the recently deceased Wuppertal zoo director Eduard Wiedemann. Heinroth wrote:

"I wouldn't mind applying for the job myself in order to get out of the Eastern zone, but on the other hand I am so deeply rooted in Zoo Berlin, and it's important that certain other forces don't gain a foothold here again. It would certainly not be for the general good, so I suppose I probably won't submit an application. Though it would, of course, be nice to be in charge of an undamaged zoo rather than the pile of rubble we have here."[53]

Incidentally, Lutz Heck also applied for the post in Wuppertal. In his CV, he claimed that he had left Zoo Berlin only after the arrival of the Red Army and the appointment of a new leader.[54] In the end, the Wuppertal job went to someone else and Heinroth remained loyal to Zoo Berlin, even turning down an offer from Leipzig.[55] By the end of 1947, she was able to show Berliners more than 570 animals of around 160 species. That was more than double the animal stock of December 1945. However, the inventory still included an awful lot of Hermann's tortoises, pigeons and other native birds. The deer reserve was particularly densely populated.[56] Apart from the lions, some bears, some monkeys and Knautschke the hippo, there were very few exotic crowd-pleasers. International animal trade or exchange was still not possible for German zoos.

In 1946, the Zoo had again been able to publish a small brochure as a kind of provisional version of its printed guide, and in 1947 almost a million visitors came to Zoo Berlin. But the next problem was already waiting in the wings. In the late summer of 1947, British troops attempted to blow up the flak tower on the Zoo's northern edge. The prior demolition of the command tower had proceeded without a hitch. Even so, on that day traffic had to be suspended throughout the whole of western Berlin. Things were different with the larger combat tower, which proved to be extremely robust and remained remarkably intact after the first demolition attempt.[57] The Zoo's grounds did not get off so lightly, however. The recently repaired roofs collapsed once more. Also, all the animals had had to be removed to a safe place beforehand. Although there were not that many animals at the time, the few staff members had been kept very busy – after all, moving a couple of lions and a hippopotamus is not like taking your cats to stay with the neighbours! Once again, Zoo employees set about repairing the animal houses and finding accommodation for the animals in their care. Then, the British Army had another attempt at bringing down the offending monstrosity. Finally, on 30 July 1948, it succeeded in breaking the tower in three using 40 tonnes of TNT. But the arduously repaired animal houses suffered yet more damage in the process. The demolition project resulted in tremendous financial damage for Zoo Berlin.[58] Once the tower had been brought down, its remains were covered with earth. One of the one-tonne bunker doors is still used as a welding table in a workshop at the Zoo.

Once all the mayhem surrounding the flak tower demolition was over, Heinroth could finally set about reconstructing her zoo. However, the next challenge – resulting from global political events – threatened the Zoo's very existence.

The expansion of the currency reform in the three Western occupation zones of Germany to the Western sectors of Berlin made things very difficult for Zoo Berlin. In July 1947, it paid its employees a bonus in the new currency, as ordered by the authorities. But the Zoo's bank account was still in the Soviet-occupied Eastern part of the city. Revenue in the new Deutschmarks from the Western sectors was insufficient to pay the employees' wages. In August 1948, the Zoo was only able to pay them ten per cent of their net salaries in the new currency and a further five per cent in the currency valid in the Eastern sector.[59] On 24 June 1948, the Soviet and East German armies and police sealed off the three Western sectors of Berlin, and the Berlin Blockade began. It was to last eleven months. The Zoo had only just begun to recover, and now it faced a fresh calamity. By that point, it was just about able to cover the amount of feed needed for its still relatively small stock of animals. Sufficient hay, litter, and meat for the few carnivores was finally being delivered from the countryside around Berlin. Now that supply route was cut off.

The Berlin Blockade from June 1948 to May 1949 posed a severe threat to the continued existence of the Zoo. Animal feed was not a priority in a city where people had to be saved from starvation via a constant stream of airborne deliveries. It was thanks to Heinroth's skills as a negotiator that the health department allocated meat from animal carcasses to the Zoo.[60] In 1948, there were still some farms within the city limits, and horses were still used on the streets as draught animals. That meant there was at least some meat available for the carnivores. Thanks to that supply, and the use of autumn leaves as a replacement for litter and hay, the Zoo's animals made it through the winter. West Berliners also demonstrated great generosity and resourcefulness to help the Zoo get through those months of deprivation. Every day they brought kitchen waste or their own food to the Zoo. They even donated hay, which was still needed for several kinds of animals in the city at that time. During the Easter week, radio station RIAS (founded by

12 December 1948
Founding of the Free Democratic Party (FDP)

Schoolchildren visit the Zoo despite the Berlin Blockade, 1948

the US occupying forces) organised a series of special events and concerts with the slogan "Save the Zoo", with all proceeds going to Zoo Berlin.

From the very beginning, Heinroth and Schröder had come up with many creative ideas to remedy the financial plight of their zoo. For example, in summer 1946 they had opened the Astra Arena close to the patched-up Antelope House. Benches and chairs from the destroyed restaurant were arranged around a small open-air circus ring, which hosted performances by what was left of Zirkus Busch, which had once occupied a vast circus building in Berlin.[61] The following year, Schröder was able to get hold of a tent. The circus performances proved to be a great success, and the Zoo received a share of the ticket proceeds. The people of Berlin were keen to take up this offer of entertainment. The fact that Heinroth and Schröder were also happy to let out their "arena" for other events such as boxing matches may have been a matter of concern for some contemporaries,[62] but Schröder had been a keen boxer during his youth and still had contacts in the boxing scene.

23 May 1949
Founding of the Federal Republic of Germany (FRG)

7 October 1949
Founding of the German Democratic Republic (GDR)

Normalisation during the Cold War, 1949–1961

Romance across the borders: Grete and Knautschke

In the early post-war years, the greatly reduced animal population of Zoo Berlin got a welcome boost thanks to a successful hippopotamus breeding programme. These efforts were a sign of the professional cooperation that still existed between zoo directors in East and West Germany – namely the old friendship between Zoo Berlin director Katharina Heinroth and her counterpart Karl Max Schneider in the East German city of Leipzig. After the Berlin Blockade came to an end, male hippo Knautschke – the only hippo to survive the war and a favourite with Berliners – received some unexpected lady visitors. Zoo Leipzig had run a successful hippo breeding programme before the war, but that was now in jeopardy following the death of its last male. Despite the increasingly frosty post-war relations between East and West, Heinroth and Schneider managed to put aside the differences of their political systems and work together across ideological and state borders. Two female hippos, Olga and Grete, were transported from Zoo Leipzig to Berlin for breeding. Grete became pregnant and eventually gave birth to the pair's first calf, a male called Schwabbel. He remained in Leipzig for future breeding purposes, while female calf Bulette, conceived during another visit, came to live in Berlin.

The numerous name ideas submitted by the people of Berlin for the city's new female hippo were highly imaginative. Some, like "Zonella" and "Eastwesta", made reference to the unusual cross-border pairing, whereas "Venus de Kilo" was a humorous nod to the hefty weight of the large semi-aquatic animals.[2] So in the end little "Bulette" got off relatively lightly – her name is a Berlin term for a type of meatball. Bulette and her father were the progenitors of a very productive breeding programme in Berlin, both going on to produce numerous offspring. Knautschke sired more than thirty calves in his lifetime. His almost legendary fertility doubtlessly contributed to the popularity of the heavyweight star, who was beloved by anyone who grew up in post-war Berlin.

Knautschke's own story is inextricably linked to that of the city. Many contemporary accounts claimed that he was born during a night-time Berlin air raid. Heinroth and her successor Heinz-Georg Klös also circulated this story – once in the

A Picasso triggerfish, 2019

A Picasso triggerfish at Aquarium Berlin, 2018

Poster advertising the Aquarium's 60th birthday, 1973

Picasso triggerfish

The Picasso triggerfish (*Rhinecanthus aculeatus*) is a solitary fish that lives in reefs, lagoons and sheltered bays. The forty different species of triggerfish inhabit a large natural range between the eastern Atlantic, the coastal waters of southern Africa, the Red Sea, and the eastern Pacific Ocean near to Hawaii. Adult Picasso triggerfish are about 20 centimetres long and very striking. They feed on algae and small invertebrates.

Picasso triggerfish become very defensive when guarding their nests. They emit crackling noises as a warning and sometimes even attack other animals and divers who get too close.

Former Aquarium director Werner Schröder was the first to bring one of these triggerfish to Berlin. The extraordinary species did not yet have a name in the German-speaking world, so art-enthusiast Schröder decided to name the beautiful fish after famous Spanish artist Pablo Picasso. He found the moniker fitting due to the fish's exaggerated triangular head and the position of its eyes, which are set unusually high up on its head. The name ended up sticking in German and even caught on in the English-speaking world.[1] In the years that followed, the fish's evocative name and image frequently adorned the Aquarium's advertising material.

Despite their beautiful appearance, Picasso triggerfish are a real challenge for aquarists. The species is very assertive and prone to aggression towards other fish. At night, the diurnal fish retreat to small caves where they anchor themselves in place using the powerful spines on their belly and dorsal fin. The latter can be locked in an erect position, giving them a secure hold and protecting them from nocturnal reef predators like moray eels and sharks. During the day, the fish's spines are an effective deterrent for hunters.

Hippos Knautschke, Jette and Bulette at Zoo Berlin, 1960

Zoo's printed guide of 1951, and again in 1983 on the occasion of Knautschke's fortieth birthday.[3] It is undisputed that he was Zoo Berlin's only surviving hippopotamus, but was he actually born in Berlin? An examination of the Zoo's old animal register casts doubt on the tale. This register records all arrivals and departures during the period in question, but the only hippo birth noted at that time was a pygmy hippopotamus. However, in October 1942, the register does list the arrival of a male hippo named Max, whose place of origin was Munich. Could Knautschke have actually been Max, a gift or loan from Munich's zoo director Heinz Heck to his colleague and brother Lutz in Berlin? Could it be that Knautschke, the "born-and-bred Berliner", was not only slightly older than people thought, but also Bavarian? That may be a bitter pill to swallow for some Berliners, but there are many indications that it is the case. After all, at that time Berlin had no breeding males that could even have sired a calf.

But does it really matter? How many citizens of the German capital can claim to be "authentic Berliners" anyway? Ever since the city was founded on the remains of a Slavic settlement, it has attracted people from far and wide, including Westphalians, Huguenots, Poles both Jewish and non-Jewish, Silesians and many more. Today, Sorbian and Polish surnames like Nowak and Kowalski, Huguenot surnames like Arendt and Behrendt, and the East Prussian surname Wuttke

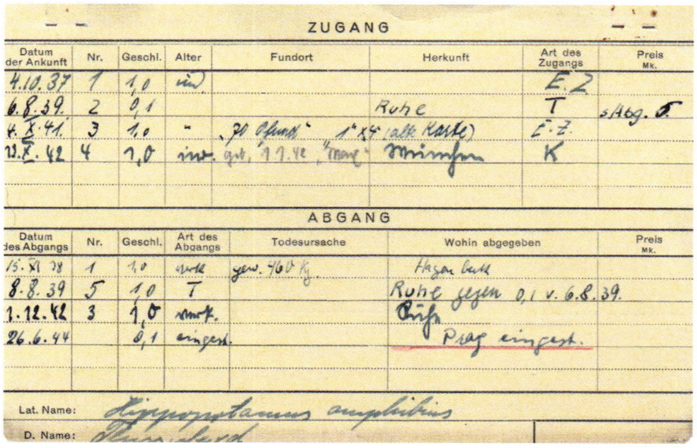

ZUGANG

Datum der Ankunft	Nr.	Geschl.	Alter	Fundort	Herkunft	Art des Zugangs	Preis Mk.
4.10.37	1	1,0	[illegible]			E.Z	
6.8.39	2	0,1			Ruhe	T	s. Abg. 5
4.X.41	3	1,0	"	[illegible]	(alte Karte)	E.Z.	
13.X.42	4	1,0	[illegible]	[illegible]	München	K	

ABGANG

Datum des Abgangs	Nr.	Geschl.	Art des Abgangs	Todesursache	Wohin abgegeben	Preis Mk.
15.XI.38	1	1,0	[illegible]	gew. 460 Kg.	Hagenbeck	
8.8.39	5	1,0	T		Ruhe gegen 0,1 v. 6.8.39.	
1.12.42	3	1,0	[illegible]		[illegible]	
26.6.44		0,1	eingest.		Prag eingest.	

Lat. Name: Hippopotamus amphibius

D. Name: [illegible]

Hippo index card, circa 1944

Cover of Zoo Berlin's printed guide, 1951

are still commonly encountered in Berlin. During the Cold War, the city attracted conscientious objectors, artists and many others from all over the world. Perhaps the "cosmopolitan" Knautschke is therefore actually a fitting favourite animal for Berlin.

A shadow of its former self

The 1950 season started strong for the Zoo with the happy announcement of the birth of brown bear triplets. The following year, another three cubs came into the world. As was customary at the time, the bear cubs had a public naming ceremony. As their sponsor was Berlin brewing company Bärenbier Brauerei, the cubs were named Schluck ("sip"), Schwips ("tipsy") and Ursator (one of the company's strong beers). In 1951, Katharina Heinroth and Werner Schröder published the first post-war report and printed guide for Zoo Berlin.[4] Naturally, celebrity hippo Knautschke featured on the front page for promotional purposes.

The map printed on the back of the guide conveys the dire state the Zoo still found itself in at that time. The animal houses are depicted with various shading: "accessible" houses and occupied facilities are dark, while "inaccessible" houses are grey. At that time, the latter still outnumbered the former. In the northeast of the grounds, the "Bunkerberg" is shown – a mountain of rubble where the flak tower once stood. The Elephant Pagoda, the Ostrich House and the Antelope House were still inaccessible ruins, as were parts of the Ape Conservatory and the Predator House. Most of the enclosures that were open could only be viewed from the outside; the indoor areas were not yet rebuilt. Only parts of the formerly so magnificent restaurant ensemble could still be used.

The printed guide also contains a historical summary of the air raids and the immediate post-

1 December 1949
Combined heat and power plant Kraftwerk West goes into operation

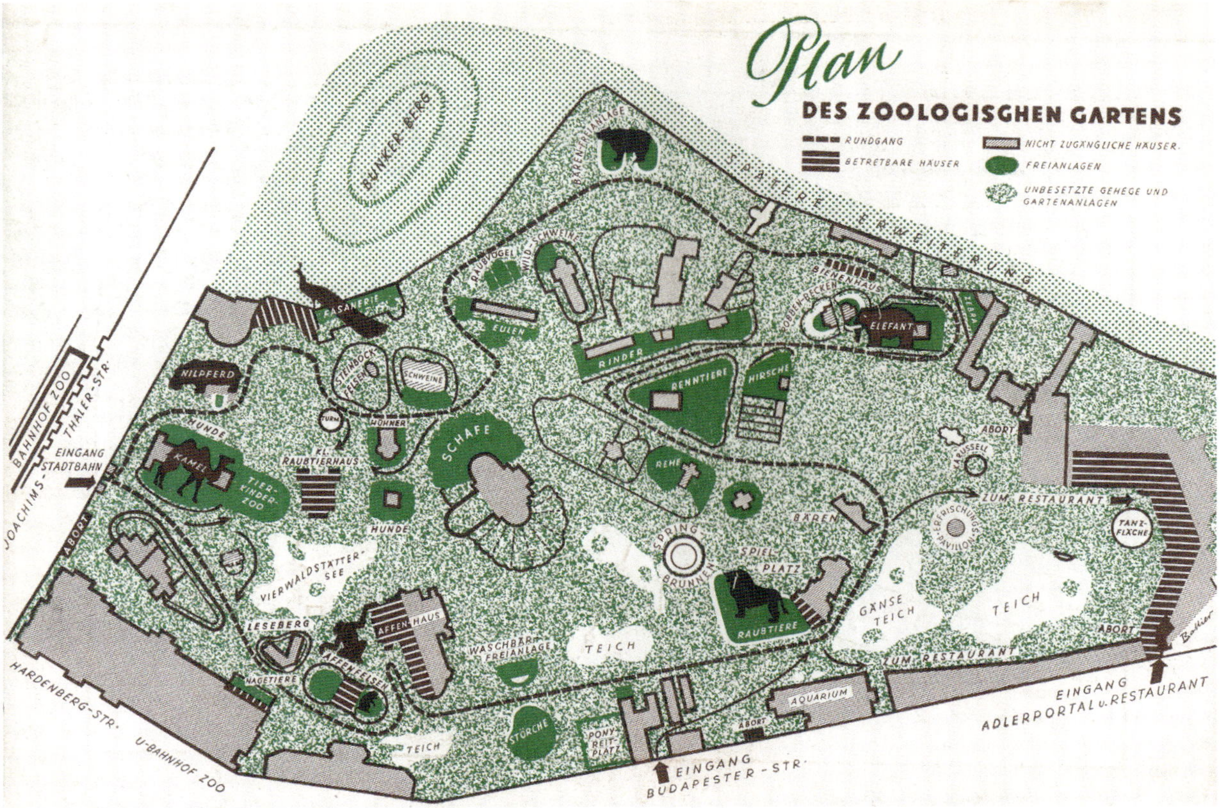

Map of Zoo Berlin, 1951

war period. It seems almost as though Heinroth and Schröder wanted to exculpate themselves by describing the damage in such detail. This allowed them to announce with a certain amount of pride that a thousand animals were now living at Zoo Berlin again.

The map also shows an elephant enclosure, even though the Zoo's only elephant to have survived the war, Siam, had passed away in 1947. Heinroth and Schröder probably kept it in out of the hope that they could soon procure another. This brand of optimism also comes across in other areas of the report. For example, the authors announced that the ruins of the Ostrich House and Elephant Pagoda were being cleared away to accommodate the expansion of Hardenbergplatz but that the Zoo was being more than amply compensated with the addition of the site of the demolished flak tower.

Shanti means peace

For many people, elephants are an essential part of any "real" zoo. On one occasion, director Heinroth received the following letter from a four-year-old fan, most likely transcribed by the child's parents: "Dear Auntie Zoo Director, Grandpa told me that your elephant is dead as a dormouse. This is very sad! Please get a new elephant very soon."[5] Of course, Heinroth knew how important it was for the Zoo to have an elephant, but she was not in the financial position to acquire one. In 1949, spirits manufacturer Carl Mampe and a private individual from Berlin offered to assist the Zoo in its endeavours. Mampe had been offered an Indian elephant by Hagenbeck, probably because the animal features in the brand's logo. The Berliner, meanwhile, had plans to contact a former schoolmate, a "Siamese prince".[6] Both those ideas came to nothing, however, so Heinroth launched an appeal via West Berlin radio station RIAS. The station's programming included school broadcasts (*Schulfunk*) and a school parliament (*Schulfunkparlament*) – an exercise in parliamentary democracy that included youth representatives from various schools in West Berlin. The school parliament wrote a letter of appeal to the Indian government on behalf of all Berlin's schoolchildren, and their campaign was successful: Prime Minister Jawaharlal Nehru complied with the request and gifted the children of Berlin a female elephant calf called Dhatri (Sanskrit for "mother Earth"). Following her arrival in Berlin on 6 July 1951, she was renamed Shanti, which means "peace". This precious gift was expressly intended for the children of Berlin – not for the city or for the Zoo. Although newly independent India had never been at war with Germany, Indian soldiers had fought on all fronts of World War II as members of the Commonwealth armed forces. Shanti was a lucky break for the Zoo and its visitors. At first, she was temporarily housed in a horse stable, where she befriended a donkey. However, her keeper and Heinroth were worried as Shanti was not sleeping. They wracked their brains for days

Shanti's naming ceremony, 1951

14 March 1950
The federal government declares West Berlin a zone of economic emergency

until Heinroth remembered that the elephant had spent her nights in India on a chain. They tried doing the same and, sure enough, Shanti's sleep improved: she was clearly not accustomed to having so much "legroom" at night.

With West Berlin's economic recovery and the introduction of the Deutschmark, Katharina Heinroth and Werner Schröder once again had modest means at their disposal to increase the Zoo's animal inventory. A companion was brought in for the approximately 20-year-old female chimpanzee Suse, who had been living alone for years. The only chimp at the Zoo to survive World War II, she was living in the still intact part of the Ape Conservatory.

Suse's keeper August Liebetreu had taught her some tricks, as was customary at the time. It is no longer considered appropriate to train chimpanzees to "ape" human behaviour, but it is possible that her training helped keep the lonely chimp entertained.

What Suse really needed, however, was a partner. Male chimpanzee Jonny was not the Zoo's first choice, but he was all it could afford. Jonny had been captured in the wild as a youngster, which required his entire family to be killed in the process – a practice that was sadly not uncommon back then. He was temporarily kept as a pet in a private household, but naturally became difficult to handle as he grew older. Jonny came to Zoo Berlin via animal trader Fockelmann in chains, which he was accustomed to. For weeks, the chimp would not let anyone near him, which meant they could not free him from his iron collar. Eventually, keeper Kurt Walter came up with a trick to liberate the chimpanzee from his chain – but Jonny by no means welcomed his new-found freedom. Instead, he spent the next few days in an angry sulk. The keepers had clearly deprived him of something he was used to. Despite this

Children sit on Shanti's back, 1955

Chimpanzee mum Suse with her baby Sam, 1954

Rhinos Meru and Arusha at the Zoo, 1955

Rieke after her return from Vienna, 1953

difficult start, Jonny ultimately settled in well and went on to sire four offspring at the Zoo, initially with Suse and then with Molli.

Jonny was a particular favourite among female visitors, as is often the case with the charismatic great apes. Jonny himself also displayed a preference for women. Zoo director Katharina Heinroth became his most trusted friend, and she made an effort to stop by his enclosure on her morning rounds. One morning, however, Jonny was extremely worked up. The director was not paying sufficient attention to notice that a noisy school class had frightened the chimp, making him uncharacteristically aggressive. Heinroth unthinkingly reached into Jonny's cage to pet him – and he bit off her finger. She was operated on in the neighbouring Franziskus hospital.[7] But Jonny was best known for his impressively accurate throwing skills. The visitors' darling was very playful and would often pull funny faces, but if he didn't like someone, he would sometimes fling anything he could get his hands on in their direction. At one point, there was even a sign at his enclosure that read: "Beware! Jonny likes to throw dirt!" Once, he even managed to knock an American officer's cap off.

By 1951, the reconstruction of the Antelope House had begun. Heinroth worked closely with the responsible architect on the plans, which not only involved restoring the building to its former glory, but also making some changes. The removal of disruptive columns in the interior, which was covered by a self-supporting domed glass roof, allowed more light into the animals' indoor area. The first famous animal to move in here was actually born in Berlin: female giraffe Rieke, who had been evacuated to Vienna during the bombing raids. In mid-August 1953, Rieke returned to Berlin – the only animal to do so of all those relocated during the war. It seems that this was thanks to Heinroth's good personal relations with Viennese zoo director Julius Brachetka, although it is true Brachetka only agreed

1 October 1950
A constitution based on the principles of freedom and democracy comes into force in West Berlin

Katharina Heinroth and Werner Schröder welcome Swampy, 1952

to return the giraffe once he was sure he would be able to make up for her loss with another acquisition.[8] Hundreds of Berliners gathered at the Zoo to welcome Rieke home and she became another important draw for visitors. Heinroth even expressed hopes that this could be the start of the Zoo recovering the rest of the approximately 1,100 animals that had been sent elsewhere – although those hopes never materialised.[9] Nevertheless, it wasn't long before visitors were able to marvel at a rhinoceros once again. The Zoo had lost its last surviving rhino to an air raid on 22 November 1943. It now welcomed three-year-old black rhino Arusha from Kenya, who it acquired thanks to financial support from the Berlin state lottery (Deutsche Klassenlotterie Berlin).

In February 1952, the Zoo also got its first post-war alligator – purely by chance. Soldiers of the United States' 6th Infantry Regiment had brought their mascot Swampy to Aquarium director Werner Schröder, as they were worried about his health. For days, the usually sprightly reptile had been sluggish and lacking in appetite. Schröder took Swampy to his heated apartment, where the Florida-born alligator made a speedy recovery. It seems he had been suffering in the cold temperatures of the infantrymen's quarters. After he got too big to stay in Schröder's apartment, Swampy was moved to the Zoo's Pheasantry, where he was given a specially prepared tank from the Aquarium.[10] The regiment realised that their mascot was better off in his new home than

17 October 1950
Elections to the East Berlin city council via a single list of candidates

A caiman in the Aquarium, 1954

Heinrich the lioness, gifted by spirits company C. K. Heinrich, 1951

Arrival of reticulated giraffe August, with the bombed-out Kaiser Wilhelm Memorial Church in the background

with them, and gifted the alligator to the Zoo – on one condition: that he would still take part in the regiment's annual parade.

As the situation in the war-torn former capital gradually started to improve, more visitor magnets joined Shanti the elephant, Knautschke the hippo, Jonny the chimp, Rieke the giraffe, and Swampy the alligator. Thanks in part to tremendous financial and political support from Bonn, the capital of the Federal Republic, West Berlin's economy slowly improved, and companies were once again able to make investments and to advertise. The Zoo's popularity with Berliners made it an attractive advertising space. What's more, the wildlife trading routes were open once again. Spirits company Heinrich, whose logo featured a

10 July 1951
Inauguration of the Berlin Airlift memorial
at Tempelhof Airport

red-maned lion, donated what they thought was a male lion. It actually turned out to be a female, who was amusingly stuck with the name Heinrich.

Department store chain DeFaKa also gifted animals. In 1955 it brought male giraffe August to Berlin as a companion for Rieke. Over the years that followed, many well-known Berlin companies followed suit. Clothing company Leineweber donated a number of animals, including two white rhinos and several gorillas. Such gifts were a popular marketing strategy as they enabled businesses to hand out free Zoo passes, which would gain them new customers. The donated animals usually came from animal dealers.

Heinroth and Schröder came up with another unusual idea to nurture the Zoo's slow-growing success: an Oktoberfest! Although the idea was met with disapproval by purists, who viewed the concept as highly inappropriate for a scientific zoo, Oktoberfests were held at the Zoo from 1950 to 1957. The success of these events justified the directors' decision, as the additional revenue helped them pay off debts. Visitors paid 25 pfennigs admission; concessions were made for East Berliners, who only had to pay one East German mark although that was worth less than 25 pfennigs based on the exchange rate. Merry-go-rounds and even a roller coaster were set up on the Zoo's grounds. The festivities lasted four to six weeks and welcomed 600,000 to one million visitors. Some years, the Oktoberfest weeks drew in more than 40 per cent of the Zoo's total annual number of visitors.[11] But smaller-scale, isolated initiatives were also effective at raising urgently needed funds for the Zoo. In June 1949, a two-hour bagpipe concert performed by the Gordon Highlanders infantry regiment of the British occupying forces attracted 12,000 visitors.[12] And in the mid-1950s, a children's railway offered a half-kilometre round trip through the rocky bear and wolf habitats. The "Zoo station" was located by the children's zoo and the train was pulled by a little locomotive with a Volkswagen engine and design, which led to it being dubbed the "VW Express".[13]

This extra revenue was desperately needed to compensate for the losses resulting from the destruction of the Zoo's eateries and peripheral buildings. The rent payments the Zoo had collected from those establishments had been a

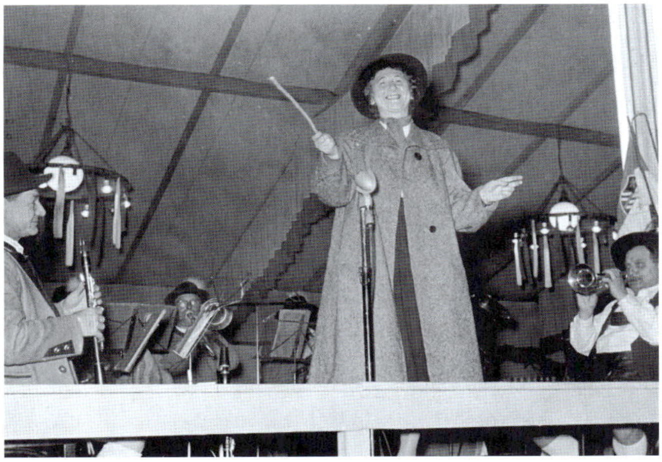

Katharina Heinroth on the stage at the Zoo's Oktoberfest, 1954

Oktoberfest poster, 1949

lucrative source of income that had long ensured its prosperity and independence from state subsidies. Without those facilities, Zoo Berlin found itself in dire straits – a situation that was certainly not helped by the fact that a second zoo in Berlin was gaining ground and becoming a genuine rival.

Competition from the East

By 1953 at the latest, serious talks were being held in East Berlin about building a zoo of its own. There were many arguments against the plan, as at that point the politically divided city had no physical barriers and people in the East were still able to visit Zoo Berlin in the Western part of the city. However, all the signs were pointing to conflict. Hundreds of thousands of people were leaving the German Democratic Republic (GDR), and no end seemed to be in sight. The violent repression of the uprising on 17 June 1953 fuelled discontent in the East, although the oppressive regime seemed stable for the time being. Within this context, the fact that East Berliners had to travel to the capitalist West to visit a zoo was a thorn in the side of the East Berlin Magistrate and the heads of the Socialist Unity Party (SED). What's more, once at the Zoo in the West, East Berliners were spending the money intended for consumption in their own republic.

When the director of Zoo Leipzig, Karl Max Schneider, was awarded the National Prize of the German Democratic Republic in Berlin in October 1953, he was taken aside during the celebrations by prime minister Otto Grotewohl and informed about the plans to set up a zoo in East Berlin. Like many zoo directors, Schneider and his scientific assistant Heinrich Dathe initially thought nothing of the project – after all, the four existing East German zoos were already fighting for their own survival. Nevertheless, the leaders of the SED wanted the capital of the GDR to have a zoo of its own.[14] Although that may sound irrational to us today, such thinking was commonplace within the logic of the Cold War. In West Berlin, politicians were no less eager to create "Western substitutes" for a number of cultural institutions and museums located in the divided city's historic centre east of the border.

In the early summer of 1954, Schneider received an invitation from the deputy mayor of East Berlin, Herbert Fechner, to come to Berlin to discuss the plans. Since Schneider was on a business trip at the time, he sent his deputy Heinrich Dathe. The Magistrate had selected three potential plots of land for the new zoo: a park in Lichtenberg, Wuhlheide public park in Köpenick, and the parkland around Friedrichsfelde manor house. Dathe's doubts about a second zoo in Berlin faded when he and a few officials visited the now-overgrown Friedrichsfelde park grounds that had been designed by Peter Joseph Lenné in 1881. According to his memoirs, he created the new animal park in his mind's eye during that visit.[15]

But who exactly was Curt Heinrich Dathe? Born in the Saxon region of Vogtland, Dathe studied zoology, botany and geography in Leipzig. In 1932, he joined the NSDAP – or Nazi Party – out of political conviction. The Nazis successfully won over young students at Leipzig University with their demands for a revision of the Treaty of Versailles and their grand promises of an economic upturn. Dathe was involved at the lowest level of the Party hierarchy, collecting membership fees in his local area.[16] From the mid-1930s, he worked as a scientific assistant at the zoo in Leipzig. He was drafted when the war broke out, but sustained a serious arm injury in France

Special guests at the opening of the Tierpark, 2 July 1955. Next to Heinrich Dathe (centre) is GDR president Wilhelm Pieck

soon afterwards. He experienced the end of the war in Allied captivity and was then allowed to return home. However, immediate reinstatement was out of the question because of his Nazi past. Dathe was able to reassume his old position in 1950.

This is where Dathe found himself, therefore, when he was asked to establish a zoological garden in East Berlin. The cornerstone of the new Tierpark – "animal park" – was laid in November 1954. The first animals to arrive were paraded through the streets of East Berlin to advertise the new attraction, as was also done in the Western part of the city. The Tierpark opened on 2 July 1955 with a great deal of propaganda and in the presence of GDR president Wilhelm Pieck. A total of 30,000 people visited the Tierpark on its opening day and it attracted more than half a million visitors in the first six months.

Numerous companies, towns and cities throughout the GDR were passionate about the project and were eager to donate money, materials and even animals. Like his West Berlin colleague Katharina Heinroth and her successor Heinz-Georg Klös, Heinrich Dathe proved to be an astute coordinator and "string-puller". Straussberg in Thuringia donated ostriches (which are called "Strausse" in German), VEB

Stamp booklet held by a participant in the NAW national construction project, 1957

Kälte – a state-owned refrigeration company – and the cooperative store in Berlin's Köpenick district gifted polar bears, and a bedroom furniture factory donated storks. It is clear that the gifts were often symbolic – and none less so than one made in autumn 1957, when the Ministry of State Security (the "Stasi") gave the Tierpark two spectacled bears. Perhaps the message behind this gift was: "We are always watching!" After all, the Stasi was not only the Party's sword and shield, it was also its eyes and ears. But above all, this gift was an opportunity for the ministry to paint itself in a positive light. The fundraising campaign led by the Stasi's paramilitary wing, the Felix Dzerzhinsky Guards Regiment, raised a considerable sum for the purchase of the bears. Dathe expressed great gratitude and invited the regiment to visit the animals.[17] Incidentally, Dathe had already acquired the bears, but now the Stasi covered the costs.

Although the motivation behind the Tierpark's planning was largely based on propaganda, that did not mean the citizens of East Berlin and the wider GDR were not enthusiastic about the project. The general public donated more than 15 million East German marks, and thousands of East Berliners volunteered to help build "their" zoo as part of the *Nationales Aufbauwerk* (NAW), the national construction project. This voluntary work programme was initiated in 1952 primarily for rubble removal and reconstruction, and it was intended to compensate for the lack of resources and machinery in the GDR at that time. Participants were given stickers for the hours they worked, which they would stick in their own personal booklets. Those who reached a certain number of hours were rewarded with bronze, silver or gold badges to pin onto their clothing.

According to the official SED party organ *Neues Deutschland*, 162,000 Berliners dedicated 690,000 working hours to the Tierpark in the period from 1954 to 1964 alone.[18] However, this form of socialist "volunteering" was highly organised and difficult to avoid. The untrained volunteers were mostly used for labour-intensive but simple excavation work. The participation of NAW volunteers in the construction of zoos was not limited to Berlin; the zoo in Rostock, for example, received 602,328 hours of free construction labour between 1956 and 1966.[19]

The press in East and West engaged in lively debate about the new animal park. While the

21 December 1952
GDR state television Deutscher Fernsehfunk goes on air

SED-controlled Eastern media celebrated various milestones in the Tierpark's construction, numerous West Berlin newspapers denounced the untenable conditions in "East Berlin's propaganda zoo", which was as much of a makeshift solution in its early years as Zoo Berlin had been in 1844.[20] Every new acquisition made by the "state-run zoo" was met with disparaging comments from the West such as, "Even the monkeys and elephants toe the Party line".[21] Meanwhile, in the East's *Neues Deutschland* paper, Dathe tried to dispel concerns about the new animal park, emphasising that the Western zoo and Eastern animal park were very different and thus complemented one another well.[22] His article echoed the words of Hinrich Lichtenstein a hundred years earlier, asking for patience and pointing out that construction at the Tierpark was still ongoing. However, due to a lack of resources in the GDR, the Tierpark was to be plagued by teething problems right up to the fall of the Berlin Wall. The West Berlin *Tagesspiegel* paper accused Dathe of "sophistry" and questioned how the recently accelerated construction efforts for the Tierpark were compatible with the "reunification rhetoric" otherwise coming from the East. The newspaper also cast doubt on the voluntary nature of the construction work and the animal donations.[23] *Die Welt,* another Western paper, took a similar stance, pointing out that the Tierpark was buying its animals in the Federal Republic, as its two polar bears were acquired from the zoo in Nuremberg.[24] Yet despite these East-West press squabbles, the zoos in Nuremberg and many other West German cities were sympathetic to their new "colleague" in the GDR and happy to help get it off the ground. Even Katharina Heinroth observed her friend's project with interest rather than scepticism.

Shortly after the opening of the Tierpark, Dathe pulled off a real coup with the arrival of

GDR citizens participating in excavation work as part of the NAW campaign at the Tierpark, 1958

Chi Chi the panda makes a break for freedom, 1958

a giant panda – exclusively in East Berlin! In the summer of 1958, the young female panda Chi Chi made a guest appearance at the new animal park. The rare bear was loaned to the Tierpark director by animal trader Heini Demmer in exchange for money – an offer that had been turned down by West Berlin. Dathe, on the other hand, agreed to pay Demmer the then considerable sum of 12,000 Deutschmarks. Largely unknown at the time was that this contract also provided Dathe with a written guarantee that Demmer would never show the panda in West Berlin.[25] Dathe invited East Berlin's mayor Friedrich Ebert Jr. to the unveiling, as well as the Western press, keen to demonstrate "how much better the Tierpark is fulfilling its cultural-political remit".[26]

Chi Chi's Berlin stopover drew 400,000 people to the Tierpark in the space of just a few weeks, after which Demmer was able to sell the panda to London. And so it was that, despite Zoo Berlin's attempts to prevent a competitor becoming established in the same city, the SED regime made the dream of a zoo in the East a reality.

Surprisingly, the arrival of a competitor in East Berlin did not lead to a drop in visitor numbers at Zoo Berlin in the West. In the Tierpark's opening year of 1955, the Zoo actually recorded an increase in visitors on the previous year. Even in subsequent years, these numbers experienced barely any decline. However, a very hard blow would be dealt with the construction of the Berlin Wall. In the first year of its existence, the notorious Wall cost the Western zoo 300,000 visitors.[27]

Finally back on track

In spite of a desperate need for new buildings following the devastation inflicted by the war and the subsequent damage caused by the flak tower's demolition, Zoo Berlin's financial dire straits meant that construction projects were not an option for a long time. Everywhere one looked there were nothing but temporary structures and makeshift repairs. If it wanted to maintain its good reputation and be more than just a local attraction, Zoo Berlin had to keep up with international trends once more – which meant that new, modern buildings were a must. Mustering up all their courage and charm, Katharina Heinroth and Werner Schröder set out to raise funds for the reconstruction and renovation of the Zoo. In 1954, the Elephant House opened as the first new post-war building. It was designed to be particularly spacious – almost as a symbol of hope for the future. And yet the number of occupants to be seen on opening day left something to be desired, with just one elephant, one rhino, and one anteater. But that situation was soon to change.

At the time, the Aquarium was one of the Zoo's biggest headaches. The building had been severely damaged in the war, but clearance of the rubble had taken a backseat as the priority was to repair the houses of surviving animals – and virtually none of the Aquarium's inhabitants had survived. On the building department's works programme of 1949, the Aquarium was listed twenty-ninth – last place – in category B, "Works yet to be carried out".[28] For a long time, it was uncertain whether it would be better to reconstruct the original building or to demolish it entirely and build something new. Eventually, the Zoo opted for reconstruction.

For this to be possible, Heinroth and go-getter Schröder once again had to obtain funds from the city government. In return for a funding pledge, they promised the Berlin Senate Department for Public Education that Berlin schoolchildren would be granted free admission to the Aquarium and they also touted the positive impact that a reopened Aquarium would have on tourism. Initial renovation efforts focused on reopening the ground floor, meaning reconstruction of the Crocodile Hall on the top two storeys had to wait until 1956. This work was eventually made possible thanks to generous financial support from the Senate and grants from the Berlin state lottery. Once the Crocodile Hall was back open and a large number of the terrariums had been restored, the Zoo Berlin management declared that Aquarium Berlin was once more Germany's foremost public aquarium, despite the fact that its restoration was not yet complete.

Visitors to the new Crocodile Hall were welcomed by the extraordinary sight of thirty of the remarkable reptiles. This was the largest collection of different crocodile species ever shown at a zoo.[29] From as early as 1953, the renovated Aquarium also housed sharks again. All areas of the Aquarium gradually reopened between 1952 and 1959. By the early 1970s at the latest, Zoo Berlin was once again home to the world's premier aquarium in terms of species diversity.

Interior view of the new Elephant House, 1955

Aquarium Berlin, 1957

Scheming and infighting

After the war, there were several years of legal dispute between the Zoo and its former director Lutz Heck about his pension rights. Heck had failed to find new employment. His application for the position of director of the zoo in Wuppertal in 1950 was unsuccessful. Due to his involvement with the Nazi Party and close relationship with Hermann Göring, the chances of him resuming a career in the tight-knit German zoological community of the post-war period seemed slim to none, despite several recommendations and testimonies from former colleagues praising the hard work he had put into their zoos and his own.[30]

A school shark in the Aquarium, 1953

Werner Schröder and guests visit the Crocodile Hall at the Aquarium, 1964

Heck repeatedly pointed out that the denazification process in the late 1940s had classified him merely as a *Mitläufer*, or a "passive follower" of the Nazis. This is hardly surprising, as given the large numbers of people that needed to be checked, the extent of an individual's guilt could often not be fully ascertained and perpetrators frequently slipped through the net. Many judges were lenient and had little interest in prosecuting. In the case of Lutz Heck, it is likely that no one believed a zoo director could have been involved in Nazi crimes. For a very long time, many regarded his role in the Nazi's nature conservation authority as unsuspicious. For Heck personally, however, true rehabilitation could only be achieved by returning to Berlin. He still had contacts in the city, his reputation here was still largely intact – and his competitor was "merely" a woman.

In 1948, he had hired a lawyer to anonymously obtain Katharina Heinroth's evaluation of his former political stance. This may have been due to the denazification processes taking place in the southern German state of Baden-Württemberg, where Heck was living. The unsuspecting Heinroth confirmed to the lawyer that Heck's "views were rooted in National Socialist ideology". She added, however, that he had been aware of her husband Oskar Heinroth's critical stance towards the Nazis but had not caused him any trouble. He had also supported the request for clemency initiated by the Heinroths for Professor Walther Arndt, who was indicted before the People's Court for criticising the National Socialist regime. However, Heinroth's own personal judgement of Heck's character was very clear: he was, she claimed, "extremely ambitious and greedy" and his friendship with Göring served to satisfy his "great ambition and thirst for power". She saw that as one of the reasons why he seized sole power at Zoo Berlin and sought to exert his claim to power over all German zoos in 1940.[31] On the other hand, the former director of Zoo Leipzig, Johannes Gebbing, wrote in 1954 to the Association of German Zoo Directors (VdZ) that Heck had helped maintain the independence of German zoos through his influence within the nature conservation department of the Reich Forestry Office.[32] While Gebbing certainly had good contacts, he had not been a zoo director since 1934.

In the spring of 1953, Heck spoke critically about Katharina Heinroth in private conversations. He claimed that a woman was not cut out to be a zoo director and that her work was unsatisfactory. At the same time, Heck began holding lectures on his animal expeditions, and he gifted the Zoo a bison hybrid from the stock at Hellabrunn. Bernhard Grzimek from the zoo in Frankfurt speculated that Heck's objective with these actions was to win favour with Zoo Berlin's supervisory board, regain popularity in Berlin, and oust Heinroth.[33] Around this time, Heck also used the media to persuade the public that his aurochs breeding programme had been a great success.[34] While his excessive devotion to the National Socialist regime had stripped him of legitimacy in the eyes of many of his colleagues, certain Berlin shareholders still saw Heck as the rightful director of Zoo Berlin.

At the 1954 annual general meeting of Zoo Berlin shareholders, the items on the agenda included, as usual, a vote of confidence in the supervisory board and the re-election of some of its members. Some of the shareholders who were old allies of Lutz Heck planned to take advantage of this opportunity; their objective was to vote out the current supervisory board and elect a new one that would bring Heck back to Berlin. One of the shareholders spoke out in favour of Heck's return in the press, accusing Heinroth of allowing too many animals to die. But there were also voices in the press that cast doubt on Heck's scientific career.[35] One of the most important of those came from Zoo supervisory board member and geneticist Hans Nachtsheim, who reiterated that Heck's prized "aurochs" were not true aurochs at all.[36] Heinroth also submitted a

7 July 1956
The West German government introduces compulsory military service; West Berlin is exempt

letter to the editor of the *Tagesspiegel* newspaper, published in the "Democratic Forum" segment, in which she emphasised once again that Heck had not left the Zoo on 2 May, as he claimed, but on the evening of 30 April, hours before the Russians invaded.[37] She probably hoped to ruin Heck's credibility in the eyes of the shareholders by showing that he had previously been caught in a lie.

However, after an hour-long meeting, the majority of shareholders still present on the night of 30 June 1954 refused to approve the supervisory board as it stood, despite Heinroth having responded to each allegation in detail at that AGM.[38] At the next general meeting in late August, all the previous members resigned and a new supervisory board was elected. Heinroth's opponents did not succeed in winning a majority, but they now had seats on the supervisory board and could continue to undermine her. It seemed that Heinroth's days as director were numbered, but she made sure to use them for the benefit of the Zoo.

Striking a good deal

Plans to make West Berlin's city centre more attractive emerged as early as 1950. The founding of two separate German nations had cemented the political but not yet physical division of the city, and politicians in West Berlin now sought to upgrade the area around the Kurfürstendamm as a replacement for the bustling boulevards they had lost in the East – namely Unter den Linden and Friedrichstrasse. After extensive planning, the decision was made to expand Hardenbergplatz and turn it into a major hub. Most importantly, Zoologischer Garten railway station, known colloquially as "Bahnhof Zoo", had become the de facto central station in the West, and the only one for long-distance travel. At a time when people were dreaming of a car-friendly city, urban planners now had their eye on the area around the Zoo. The proposal put forward by some planners to build a helipad on the path alongside the Landwehr Canal, which would enable West Berliners to reach West German cities like Hanover or Hamburg in just over an hour, remained nothing more than an anecdote.[39] A more sensible concept was the creation of a central bus station and large car park in front of Bahnhof Zoo, which had previously only been separated from Zoo Berlin by a narrow road. Implementing this plan would require the Zoo to give up about 60,000 square metres of space. The structures that stood on this plot of land at the time were the ruins of the Elephant Pagoda, the Ostrich House and the Hippo House. The Lion's Gate entrance was also to be moved several hundred metres, and the once-magnificent buildings that now stood derelict on the Zoo's Budapester Strasse periphery were to be knocked down to allow for a wider street and new shops.

The Zoo was compensated for these losses with approximately 90,000 square metres of land to the west of the railway tracks, to be used for administrative and utility purposes, and with the site of the former flak tower as far as the Landwehr Canal. The Senate was also prepared to compensate the Zoo with an annual subsidy for the loss of rental income from its now permanently removed peripheral buildings. The Zoo even received financial compensation for the remains of the Elephant Pagoda that were lost through the exchange of land, on the condition that the money would be used to build a new elephant house.[40] Funds were also provided for the removal of 412,000 cubic metres of rubble from the demolished flak tower. In the 1960s,

Katharina Heinroth opens the topping-out ceremony for the Elephant House, 1954

the first of a number of increasingly modern bird houses was built on this site.

All those decisions showed just how much the West Berlin Senate cared about the Zoo. After all, the resulting situation was very advantageous for the Zoo, which would have struggled financially to renovate its peripheral buildings by itself. However, it can also be assumed that the Senate's generosity was motivated in part by a desire to increase and secure the head start that "its" zoo already had over the new competitor in the East.

Negotiating this very beneficial deal was Katharina Heinroth's last gift to the Zoo as director. As with so many of her previous achievements, Heinroth did not even get a thank you from the supervisory board – now an old boys' club whose members wanted to hand management of the Zoo back to Lutz Heck. According to Heinroth, board member Fritz Schmidt-Hoensdorf approached her in the autumn of 1955 and announced that they had decided to replace her with a man. He conceded that she had done a good job, but claimed that a woman could not be expected to oversee the required construction efforts.[41] The official reason given later was that the board was looking for a younger director. Schmidt-Hoensdorf was a former zoo director from Halle, but he had not returned to his position after 1945. Perhaps due to the fact he had been a member of the Nazi Party, the National Socialist German Lecturers' League and an SS equestrian unit,[42] he opted to stay in West Berlin during the post-war period, where he worked as a businessman and lecturer. He may have hoped that by reinstating Heck, he too could resume directorship of a zoo.

26 November 1956
Opening of the first section of West Berlin's inner-city motorway

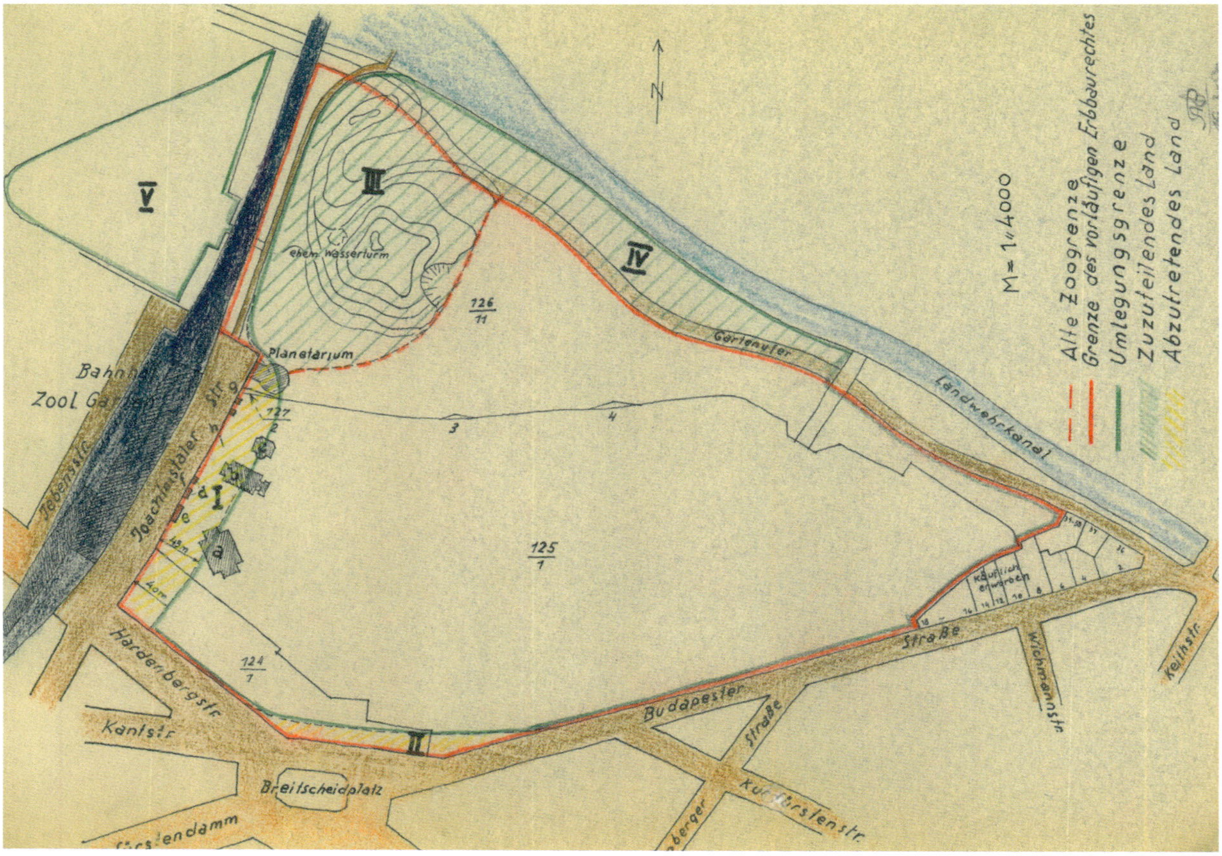

Map showing plans for land changes around the Zoo, 1955

But Heinroth had her own share of friends and admirers. Prior to the annual general meeting, the Zoo received a letter from a female shareholder calling on the supervisory board to explain its reasons for Heinroth's dismissal. It fell to Heinroth herself to forward the letter to the board.[43]

Although she was not able to take any real action against her forced retirement, Heinroth did not intend to give up without a fight. At her last general meeting as scientific director on 27 June 1956, she addressed the dismissal as part of her review of the previous year, complaining that the supervisory board had refused her an assistant for years and yet was now putting forward the argument that they needed someone younger at the helm.[44] When the chairman of the board Theodor Frenzel and other board members started to heckle her loudly in an attempt to disrupt her speech, the shareholders intervened. With rounds of applause that lasted for minutes, they saw to it that Heinroth was able to speak and gave the zoo director, who was very popular with the public and shareholders alike, the respect she deserved for her many achievements.[45] Heinroth's final legacy was a business year that had seen an increase in visitor numbers and revenue. Also, Zoo Berlin was now home to far more species than at the end of the war. Thanks to Heinroth's tireless reconstruction work in collaboration with Werner Schröder and her perseverance through many difficult years, the Zoo's surviving population of ninety-one animals in May 1945 had now grown to almost 2,000 animals, representing some 470 species and subspecies.[46] Zoo Berlin was still far from regaining its position as one of Europe's leading zoos, but given the extent of the wartime damage and the opposition that she had faced, Heinroth had done an outstanding job – particularly as she had initially only assumed the role out of a sense of duty towards her late husband, Oskar Heinroth. It would be up to her successor to continue the good work and turn the facility into a modern zoo. Katharina Heinroth felt very strongly that the main priorities should be the construction of a new house for the apes and monkeys and the rebuilding of the Predator House.

In the end, the supervisory board also granted Heinroth the right to propose a successor. The names she put forward included Heinrich Dathe of the Tierpark in East Berlin, Frankfurt Zoo director Bernhard Grzimek, Cologne Zoo director Wilhelm Windecker and, in fourth place, a young colleague from Osnabrück named Heinz-Georg Klös. After the first three politely declined, Klös was selected.

A "whippersnapper" takes the helm

And so the task of completing the modernisation of Zoo Berlin fell to Germany's youngest zoo director of the day. Katharina Heinroth was not displeased with the choice of successor, but the handover date forced upon her by the supervisory board left her with no time to properly train the new director.[47] Almost as a final act of humiliation, the supervisory board decided that Heinroth must take her remaining annual leave in the last months of the year before her contract ended. The new director was then to assume his position on 1 January 1957. Heinroth protested this decision, stating that it was important for the director to be present when the first frosts hit, both for the animals and for the construction projects. Above all, however, this would leave her no time to show Klös the ropes. She explicitly refused to accept responsibility for the consequences of that decision.[48] But the supervisory board insisted.

Klös later reported that it was, in fact, board member Schmidt-Hoensdorf who persuaded him to come to Berlin. Perhaps the "old boys" assumed he would be easy to control. Given Klös's tender years and relative lack of experience, some Berliners and Zoo staff were initially sceptical about the "new chap". In a letter to the Zoo prior to the 1956 annual general meeting, one shareholder pointed out that Klös was still an untrained animal keeper when Katharina Heinroth had begun rebuilding the Zoo in 1945.[49] Some even believed he must have been a relative of Schmidt-Hoensdorf's, given how strongly the latter had campaigned for his appointment.[50]

Heinz-Georg Klös was only 30 years old when he assumed his new role in January 1957. However, he already had many years of zookeeping experience under his belt. He had grown up near the zoo in Wuppertal, where he had gained work experience at the age of 14. He consolidated this experience with internships at various zoos and the circus. After being released from British captivity following the war, he studied veterinary medicine at Justus Liebig University in Giessen from 1947 to 1953. In April 1954, Klös became the youngest zoo director in Germany at the zoo in Osnabrück.

In Berlin, Klös faced a number of challenges and widespread mistrust. The forced ousting of his popular predecessor meant that many did not welcome his arrival. The new director also inherited a difficult court case: Lutz Heck was suing his former employer for pension payments. The Zoo's supervisory board had refused these payments, claiming that the board's disempowerment as a result of Göring's decree in the spring of 1944 had been initiated by Heck and thus constituted a breach of loyalty, not to mention the fact that Heck fled from Berlin and subsequently had funds transferred to southern Germany. The Berlin regional court had initially found in favour of the supervisory board, but the superior court ruled that Heck's claims were justified.[51] The case was thus set to continue.

At the same time, however, the Zoo received a financial windfall. In the spring of 1960, it was granted the large sum of 500,000 Deutschmarks from the ever-supportive lottery fund as part of its grant distribution programme. This money was immediately invested in constructing a bird house with a special parrot section. Shortly afterwards, in April 1960, the West Berlin Senate pledged 15 million Deutschmarks for the clearance of all wartime ruins and rubble at the Zoo – on the condition that it be granted three of the twelve positions on the supervisory board so that it could control the funds. At the shareholders' meeting, the statute amendment was approved and new members were elected.[52] Support like this from the West Berlin city government was naturally welcomed with open arms by the Zoo, but it came at a price: greater dependence on politicians. In the future, striking the right balance between receiving state support and retaining control would be a tricky task for the Zoo

Visitors enjoy the sunshine by the Monkey Rock, 1957

July 1958
The fifth SED Party Congress decides to intensify and collectivise agriculture

November 1958
Beginning of the "Second Berlin Crisis" over the city's political status

Zoo Berlin's new young director Heinz-Georg Klös with Walt Disney, 1958

Interior view of the new Primate House, 1958

management. For the time being, however, Klös focused on continuing his predecessor's efforts to gradually build up the Zoo's animal inventory.

A race for supremacy

In order to survive, zoos have to attract visitors. But if a zoo is not popular with the public, it will lack the financial resources required to construct more appealing buildings and acquire interesting species. This vicious circle can sometimes be broken through generous donations. But mostly, zoos rely on government grants. Until 1950, Zoo Berlin was only able to increase its still rather meagre animal inventory thanks to donations from individuals, as commercial purchases were not possible because of difficulties relating to the currency reform. Once the situation allowed, therefore, Katharina Heinroth and especially Heinz-Georg Klös were very active in the purchase of new animals. Heinroth embarked on an expedition to East Asia in 1955, from which she returned with twenty new animals, including two orangutans and a young pig-tailed macaque. These animals formed the beginnings of a new primate collection, which Klös then had to house.

Klös was an advocate for systematic zoos, and confessed to having a personal preference for "rarities as well as exotic show animals".[53] His scientific ambitions often had to take second place in order for him to increase the attractiveness of the Zoo. Klös knew he could only afford to purchase the species that were of most interest to him if he acquired animals that were particularly popular with the public. The construction of a new house for the apes and monkeys was therefore, as Heinroth had recommended, at the top of Klös's agenda. Across four construction phases, the new Primate House was built on the site of the former Ape Conservatory. The money for that structure, which was completed in 1965, came from Berlin's housing loan institute, the

2 December 1958
Opening of the Hilton Hotel
on Budapester Strasse

18 June 1959
Schloss Bellevue becomes the official Berlin residence
of the federal president

Wohnungsbaukreditanstalt. Unlike the previous house, the new facility was to include spacious outdoor habitats. Another new feature inside the house was the separation of visitors and animals by panes of glass rather than just bars to prevent the transmission of disease.

The situation in East Berlin was very similar at that time, with the Tierpark also keen to acquire a stock of animals that would draw in visitors. So far, high visitor numbers had provided clear confirmation of the excellent work carried out by Heinrich Dathe and his colleagues. In 1956, only its second year of existence, the Tierpark's visitor numbers almost hit the one-million mark. It would remain among the top German zoos in terms of visitor numbers for several years to come. In 1959, more than 1.5 million people visited the Tierpark. In the previous year, the droves of visitors included many West Berliners, who were transported to the new animal park in the East by the Berolina coach company. These organised trips were eventually prohibited by the West Berlin administration, a clear sign of how popular the attraction was.[54] The Tierpark advertised with large posters at all local railway stations in the city – including Zoologischer Garten.[55] This was possible as a result of Allied regulations that had the East German railway company Deutsche Reichsbahn operating the local light rail service in both East and West Berlin until the 1980s.

Nevertheless, Dathe found himself repeatedly having to court GDR officials or force their hand with gentle pressure – usually involving Zoo Berlin in the West. In 1960, construction work at the Tierpark ground to a halt; after initial enthusiasm, the East Berlin Magistrate's interest in this prestige project had stalled. Already in the summer of 1956, Heinrich Dathe had contacted Deputy Prime Minister Walter Ulbricht to urgently request a delivery of 1,600 tonnes of cement – with little success.[56] In early 1958, Dathe warned that the ongoing makeshift nature of the facilities was leading to a variety of complaints, endangering the Tierpark's good reputation thus far with foreign and West German visitors. Given the efforts being made by Zoo Berlin in the West to catch up, he said, the heated animal houses urgently needed to be completed.[57] In 1960, he approached the Magistrate's department of culture and cautioned that he would soon be forced to sell a number of show-worthy specimens if permanent animal houses were not built. In fact, Dathe even threatened to sell up to 60 per cent of the Tierpark's stock if the building plans were reduced any further. The safety of visitors and animals, he explained, could not be guaranteed with the temporary facilities.[58]

In April 1960, Dathe talked with East Berlin's councillor for culture Johanna Blecha, painting a bleak scenario of "mass public migration" to the Zoo in the West. He claimed that Heinz-Georg Klös would soon be receiving a delivery of twenty-two chimpanzees. Dathe demanded more money and a long-term plan, similar to the one developed for Zoo Leipzig.[59] Whether Dathe actually believed that Zoo Berlin would receive twenty-two chimpanzees at once is difficult to say. Construction work on the Primate House at Zoo Berlin had certainly not progressed enough to accommodate such a delivery. He was probably just trying to put serious pressure on the Magistrate, disappointed that its early enthusiasm had faded.

The twenty-two chimpanzees never did materialise, but the construction of the Primate House in the West remained a sore point for Dathe. In 1957, clothing company Leineweber had donated a young male gorilla named Knorke to the Zoo. Gorillas have always been favourites with zoo visitors. Knorke contracted paratyphoid fever

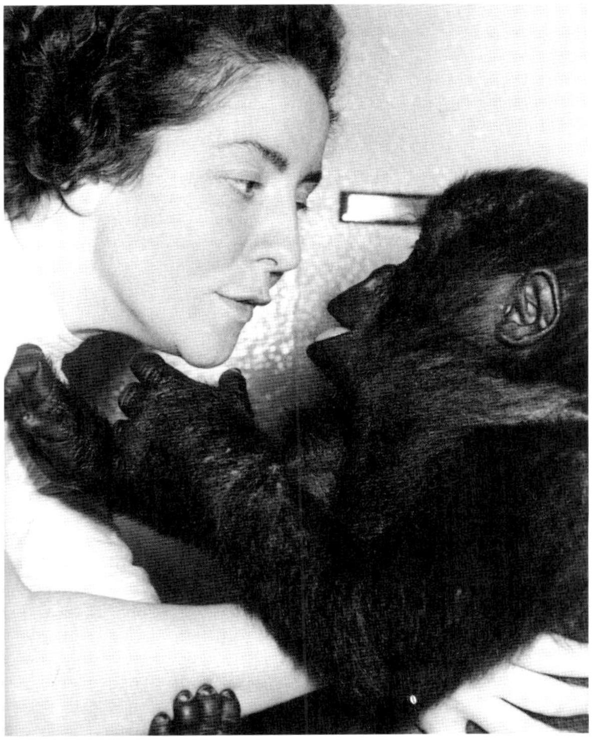

Knorke and his nurse Rosemarie Hohler, circa 1957

Crowds flock to see Knorke after his arrival at the Zoo, 1958

and was assigned his own paediatric nurse from the Moabit municipal hospital, as the disease patterns of children and young apes are similar. From June 1957 to July 1958, nurse Rosemarie Hohler was released from her usual hospital duties in order to look after her small, hairy patient in a specially set up area of the clinic.[60] She kept a detailed journal of Knorke's condition and development during his time in her care. Heribert Pauly, the senior physician of the paediatric department, was responsible for the ape's medical treatment. This unusual situation produced a lot of emotionally-charged press reports – the best publicity Zoo Berlin could have hoped for. In 1959, female gorilla Fatou arrived, followed by Gigi in 1961. The Zoo's increasing number of great apes, which are so fascinating for visitors, represented a real threat to the Tierpark.

But Klös also had to be mindful of the competition from the East in his plans, and he was just as preoccupied with how he could improve the standing of Zoo Berlin. In 1959, the Zoo found itself in tough negotiations with Berlin politicians. Working with Aquarium director Werner Schröder and Zoo Berlin's new commercial director, Klös had put forward big plans that required 35 million Deutschmarks in funding. The management team wanted to invest in new buildings and the full reconstruction of the Zoo over the next few years. Although the Zoo was able to sustain itself financially thanks to the compensation it had received from the land exchange deal with the Senate, it did not have sufficient capital to develop the two new areas of land that it had acquired on the site of the former flak tower and beyond the railway tracks. And what argument could be more convincing than one involving a battle for supremacy between the two political systems? In these negotiations, the Zoo's management emphasised that the threat the Tierpark posed should not be underestimated, since its outdoor enclosures, including the polar bear

Knorke greets Fatou on the runway of Tempelhof Airport, 11 May 1959

outdoor habitat, had been built with the strong support of the government of the "Soviet zone", which spared no expense to make the Tierpark "an institution of a highly significant nature". But the Zoo and its Aquarium, they claimed, represented an important link between the people of West Berlin and the Federal Republic of Germany, and their "brothers and sisters" that came from the Eastern zone and Eastern sector.[61]

West Berlin's politicians were already sympathetic towards the Zoo, as it played a part in West Berlin's depiction as a "showcase for freedom", and Klös found an advocate in the senator for public education. In light of the "serious competition" coming from the Tierpark, he believed that the Zoo needed an area of specialisation. The senator agreed that the ape and monkey enclosures and their stock should be expanded, as he believed this was an area the East was unlikely to focus on due to its iron shortage and the foreign exchange situation. Instead, he claimed, the East would probably invest its energies in outdoor facilities, where experience had shown that the difficulties surrounding material availability didn't matter nearly as much.[62]

Later, when Klös was seeking money for new

Ulrich Nagel, caricatures of Heinz-Georg Klös and Heinrich Dathe, 1958

bear enclosures in the late 1960s, his funding requests were accompanied by observations that the Tierpark had built a large outdoor bear facility and "it is now the task of Zoo Berlin to build one that is equally beautiful, if not more so".[63]

This sort of Cold War rhetoric played a role for a long time in divided Berlin, including at its zoos. For example, in 1983 Klös was very keen for West Berlin's mayor Richard von Weizsäcker to consistently refer to the "Zoo Aquarium" during his speech on the occasion of the reopening of the renovated Aquarium. He knew that the Tierpark management always presented the Zoo and Aquarium as separate entities so that it could claim to have more animals than the Zoo.[64]

There has been much speculation about the two directors and their relationship to one another. Heinz-Georg Klös and Heinrich Dathe shaped their respective zoos for thirty-five years and both became prominent public figures, making regular appearances on radio and television and being honoured with multiple awards. They presided over locally and internationally competing institutions, which were very popular with Berliners in both halves of the city. While their rivalry meant the two men were certainly not friends, they did develop a mutual respect for one another's achievements. This was not open hostility, but healthy, spirited competition between two zoologists who both had a good sense of the political mood and knew how to make use of that to benefit their respective institutions.

Heinz-Georg Klös and Heinrich Dathe, 1984

A zoo on an "island": The 1960s

The Berlin Wall goes up

Following a difficult restart after the war, by the mid-1950s Zoo Berlin was once again functioning properly as a modern zoo. In late April 1960, it introduced a discount for visitors from the GDR. On "low-cost Sundays", Children's Day and public holidays, Easterners could receive a food voucher worth 80 pfennigs for just one East German mark – a good discount compared to the official exchange rate. The initiative was intended as "a modest but clear expression" of the Zoo's sense of unity with its "compatriots on the other side".[1] The fact that many East Berliners and Brandenburgers still visited the Western Zoo despite the competition in Friedrichsfelde is clearly demonstrated by the abrupt decline in visitor numbers brought about by the construction of the Berlin Wall in 1961. In the year following this event, the Zoo had 300,000 fewer visitors.[2]

Acting on the orders of First Secretary of the ruling Socialist Unity Party Walter Ulbricht, GDR police and soldiers effectively closed the border between East and West Berlin with barbed wire and anti-tank obstacles on 13 August 1961. Over the days that followed, they began constructing a wall more than two metres high. The already politically divided city was now also physically split, with the Berlin Wall severing roads and other transport connections and brutally separating loved ones. Windows facing the Wall were bricked up, and those who attempted to flee were shot at or imprisoned for a long time.

The physical division of the city also had an impact on the Zoo, which now needed to heighten its appeal for West Berliners and West Germans. The former was quite easy to achieve: as West Berlin was now walled in, inner-city woodland, lakes and green spaces had become a valuable replacement for the countryside the people could no longer access. The Zoo was an important green space; a piece of nature right on their doorstep. The Senate was asked to explicitly include the Zoo in its recommendations and official travel programmes for West German visitors to Berlin.[3] Already by the following year, the Zoo recorded roughly the same number of visitors as it had in 1960.

But West Berliners had not forgotten entirely about the Tierpark in the East. That is the only explanation for the slight drop in visitor numbers that the Zoo experienced for a few years in the

A bald eagle, 2018

Bald eagle

The bald eagle (*Haliaeetus leucocephalus*) is one of the best-known heraldic animals in the world, as it adorns the Great Seal of the United States of America. But even before the country was established, the bald eagle had long held significance for the land's Indigenous peoples. The bird remains a significant motif in Native American art to this day, and its feathers are used in religious rituals.

The bald eagle's natural range covers more than just the US mainland; it can also be found in Canada, the Aleutian Islands, parts of northwestern Mexico, and the US Virgin Islands. The birds tend to be found by the ocean or other large bodies of water. Bald eagles are very adaptable when it comes to diet; they are happy to eat fish, other birds, small mammals, reptiles and invertebrates. They even feed on carrion. Their large wingspan of more than two metres makes bald eagles less swift and nimble in the air than smaller raptors, but they occasionally steal prey from other birds.

Bald eagles reach sexual maturity at around four years of age. Females lay two to four eggs, which are then incubated for around five weeks in nests, or eyries, that are often more than two metres wide. Once hatched, the chicks spend almost three months being tended to in the nest until they are able to fly. The bald eagle's courtship dance is particularly impressive: males and females fly towards each other at high altitudes, lock talons and start to tumble out of the sky. When they are within just a few metres of the ground, they break free of this "embrace" and fly back up to repeat the ritual.

Despite having no natural enemies in the wild, bald eagles' numbers in the United States started to fall from the late 1700s onwards as a result of hunting.

A bald eagle

The population then plummeted with the increased use of the pesticide DDT, which the birds ingested with their prey. The chemicals in DDT greatly reduced the quality of the eagle's eggshells, resulting in fewer offspring. Thanks to conservation measures, today the species as a whole is no longer considered endangered. But individual populations are still in need of protection.

Zoo Berlin has housed bald eagles since 1906. The most famous was a bird the Berliners dubbed "Willy Brandt", who arrived in West Berlin after the construction of the Berlin Wall as a gift from US Attorney General Robert F. Kennedy.

early 1970s.[4] By the end of 1973, the Four-Power Agreement on Berlin had entered into force, and West Berliners could now travel much more easily to East Berlin, at least for day trips. In addition to visiting family, it is likely that many West Berliners used the opportunity to visit the Tierpark. However, visitor numbers at the Zoo in the West began to rise steadily again after 1973.

The Cold War and West Berlin's situation as an "island" in the middle of the GDR led to the divided city boasting a rather unusual collection of animals. Politically motivated gifts to zoos have a long history. That was, after all, how Zoo Berlin was originally founded – with the Prussian kings

Vietnam's head of state Ho Chi Minh visits the Tierpark, 1957

Kosko's welcoming ceremony at the Tierpark, led by Heinrich Dathe, 5 May 1958

6 June 1961
Founding of ZDF, a second national TV broadcaster in West Germany

15 June 1961
Walter Ulbricht makes the infamous statement "Nobody has any intention of building a wall"

Kosko with Falk Dathe – son of the Tierpark director and later a zoological curator, 1958

donating their animal stock from Peacock Island and with extravagant gifts from foreign rulers. Zoos are an excellent stage that rulers and political systems can use to represent their country via exotic animals.

It stood to reason, therefore, that the Tierpark in East Berlin would also receive such gifts – especially from Eastern Europe and other Socialist countries. In 1956, the Soviet Union gifted the eastern zoo a Siberian tiger, and in 1957, the zoo in Beijing gave it a pair of Chinese alligators. The following year, a larger shipment arrived from North Vietnam, which had recently gained independence. Vietnam had emerged from the First Indochina War as a divided nation, and the northern half was now under Communist rule. Its head of state Ho Chi Minh visited East Berlin in 1958. He enjoyed an extensive tour of the Tierpark with Heinrich Dathe, who was able to communicate with him in French. As a sign of solidarity to its "brother state", but certainly also as a political gesture, that year North Vietnam gifted Tierpark Berlin its first Vietnamese pot-bellied pigs and a young elephant. The pigs became the progenitors of a successful breeding programme while the two-year-old elephant calf Kosko roamed the grounds freely, much to the delight of visiting children.

The US government clearly did not want to be outshone by the Communists. On 23 February 1962, US Attorney General Robert ("Bobby") Kennedy presented the Zoo with a bald eagle – the national bird of the United States. The event saw thousands of Berliners flock to the Zoo. Coming just half a year after the erection of the Berlin Wall, Kennedy's visit and his symbolic gift were interpreted as a sign of close friendship with the protective power. Kennedy was only scheduled to visit the Zoo for 15 minutes, but the rush of enthusiastic Berliners eager to welcome this representative of the "leader of the free world" was so great that his visit took much longer, with Kennedy and Klös confined to the relative safety of the eagle aviary for quite some time. Unfortunately, the gifted eagle was not quite as strong the public's enthusiasm. Just two years later, the rather feeble bird died and a replacement had to be acquired.[5]

The bond between West Berliners and Allied forces had been growing ever closer since the days of the Berlin Airlift. The former occupiers had become friends, and Berliners were keen to express their gratitude. In 1957, the Zoo had loaned one of its own heraldic animals – a young bear – to the United States' 6th Infantry

Regiment as a mascot. In 1958, the animal was returned to the Zoo, with a military ceremony.[6] The traditional mascot of the Royal Welsh Fusiliers, a male goat named Billy, received the same honour. Billy was handed back to the Zoo, which had been in the British sector, in 1992 when the Allied forces withdrew.

After the Wall went up, the Tierpark in the East lost its local rival. The nearest large, established zoos were now in Leipzig and Dresden. That undoubtedly heightened the significance of the animal park in the GDR's capital. Despite the supply shortages in the GDR, Dathe and his staff continued to develop the Tierpark tirelessly. Sometimes they lacked even basic resources. For example, the GDR Ministry for Culture was not even able to obtain the standard ornithological work, *Handbook of the Birds of Central Europe*.[7] However, in 1963 the Tierpark did gain an impressive new animal house for its big cats – the

Robert Kennedy at Zoo Berlin, 1962

The Welsh Fusiliers return Billy to the Zoo, 15 May 1992

19 to 21 August 1961
US Vice President Lyndon B. Johnson comes to Berlin

17 August 1962
Eighteen-year-old Peter Fechter bleeds to death after attempting to cross the Wall

October 1962
Cuban Missile Crisis

The Tierpark's Alfred Brehm building, 1965

Alfred Brehm building, designed by architect Heinz Graffunder. The building was considered exemplary from both an architectural and animal-keeping perspective and it achieved global recognition from other zoos and zoologists.

One of the best-known zoologists of the day, Heini Hediger from Zurich, congratulated Dathe on this "landmark in the international history of zookeeping".[8] The ultra-modern central glass roof covered a tropical hall, with cages for exotic birds and enclosures for predators arranged along the building's long sides. The two open rocky enclosures at opposite ends of the building were a particular sensation as the first-ever open indoor enclosures for big cats. Set behind wide moats, these rocky habitats did not necessarily reflect the natural homes of their inhabitants, but they certainly provided a fitting stage for the magnificent beasts. Dathe allegedly postponed the opening of the Alfred Brehm building until the seventieth birthday of Walter Ulbricht. Perhaps Dathe's strategic attempt at flattery was successful.[9] The Alfred Brehm building is now a listed building – a fact that did, however, pose particular challenges during renovation work on the animal house to bring it in line with modern standards.

The Zoo turns 125

West Berlin's situation as an "island" inside East Germany ultimately attracted many visitors to the walled-in city and to Zoo Berlin, which also provided a fitting backdrop for a variety of special events. In September 1967, for example, the West German national football team posed for pictures in front of the Zoo's lion enclosure. The team then went out to beat France 5:0 in the sold-out Olympic Stadium.

In the early 1960s, the Zoo embarked on its largest post-war construction project to date: the rocky outdoor bear habitats. Klös had already promised the construction of a large area for bears in his first annual report. Since the bear is Berlin's heraldic animal and features on its coat of arms, Klös wanted the Zoo to achieve "as complete a bear collection as possible".[10] He planned spacious rocky enclosures for all the Zoo's bear species. At the heart of this new area was the existing brown bear enclosure that had been created by Lutz Heck for his "German zoo". Soon, the enclosures for polar bears, brown bears, tropical bears and wolves were finished. These were arranged around an area hidden from view from visitors that was intended for the keepers and background operations. Space for the new attraction had been freed up following clearance of the rubble of the former flak tower. Ground was first

Members of the West German national football team with a young lion at the Zoo, 1967

15 October 1963
Opening of the new Philharmonie building

16 October 1963
Ludwig Erhard becomes federal chancellor

Opening ceremony for Zoo Berlin's 125th birthday celebrations in the West Berlin Kongresshalle, 1969

A poster advertising the Zoo as an "oasis in the global city", circa 1964

broken in March 1966, and by the start of 1969 a total of nine outdoor enclosures had been completed, as well as a smaller enclosure for the rearing of polar bear cubs. Step by step, occupants moved into each facility and were presented to the public. The first was the 1,400 square metre polar bear enclosure with its large moat that doubled as a swimming pool for the bears. The company Reul Granit had donated 2,100 tonnes of granite from its Niederlamitz plant, which had been transported to Berlin on 200 railway wagons.[11] Once in the city, it was brought to its final location with the help of soldiers from the British Army's Royal Engineers, as the Zoo was situated in the British sector. The occupying forces, which had become a protective power for the West Berliners, were always happy to help out free of charge, for example by lending heavy equipment and cranes. Six polar bears from Russia (one male and five females) were donated by the long-standing Berlin furniture company Möbel Hübner. Later, four more young bears were added to the Zoo's stock.[12]

Since it had not been possible to properly mark the Zoo's hundredth anniversary, there was an even greater desire for its 125th birthday to be celebrated in style. It was 1969 and things were looking up for the people of West Berlin. The "economic miracle" was still proceeding apace, and a series of travel permit agreements meant that West Berliners were able once more to visit their relatives in the East. With perfect summer weather to boot, the anniversary event was a huge success. In this same year, Zoo Berlin hosted the annual meeting of the Association of German Zoo Directors (VdZ). The meeting was timed so as to enable the zoo directors to participate in both the official anniversary celebrations and the fair for the general public. Klös was thus able to show his counterparts at other German zoos the

1 December 1966
Kurt Georg Kiesinger
becomes federal chancellor

2 June 1967
Student Benno Ohnesorg is murdered
by a police officer

February 1968
First student protests in West
Berlin against the Vietnam War

good connections that Zoo Berlin enjoyed with the city and its highest representatives, as well as the great affection that the people of West Berlin had for their zoo.

The anniversary celebrations were accompanied by painting competitions for schoolchildren as well as the issuance of special stamps. Events began on 2 July 1969 with a children's festival at the famous Kaufhaus des Westens (KaDeWe), or "Department Store of the West". The following evening, the grand opening ceremony took place in the Kongresshalle conference hall with 1,200 guests, introduced by West Berlin's Governing Mayor Klaus Schütz. Musical accompaniment was provided by the horn band of the Berlin hunting association and the military band of the Berlin police, who performed hunting fanfares and *The Poacher* by Albert Lortzing. A film by renowned wildlife documentary maker Heinz Sielmann was also screened. On 5 July 1969, the Zoo invited all Berliners to its big birthday party in the zoological garden itself. Anyone who shared their birthday with the Zoo was granted free admission.

Local radio station Sender Freies Berlin and TV channel ARD broadcast two hours of the event in the afternoon, with performances by well-known Schlager singers like Nana Gualdi, Paul Kuhn and Rosi Rohr as well as Berlin political satire troupe Die Stachelschweine.[13] A treat for the eyes and the tastebuds was the 125-metre-long cake, which was devoured by 40,000 guests – proof of the popularity the Zoo had once again attained. Not only did the Zoo receive a generous donation of 600,000 Deutschmarks as a birthday gift from the West Berlin Senate, income also poured in from the sale of special postage stamps and commemorative coins, as well as from numerous special donations and increased ticket sales.[14]

However, the high spirits of the anniversary year were dampened somewhat by the fact that, in January, GDR zoo directors had withdrawn from the aforementioned VdZ, which had continued to exist after the founding of two German states. Klös, who was then president of the association, received one resignation letter after another. Some offered little by way of explanation, while others like Tierpark director Dathe cited the "increasing tensions" between the Federal Republic and the GDR as the reason for their departure. Zoo Leipzig director Siegfried Seifert claimed his resignation was a result of the Federal Republic's claim to sole legitimacy. The Cold War had thus infiltrated the German zoo community. Reading between the lines, it appears that not everyone left voluntarily. Rostock Zoo director Dieter Schwarz assured Klös that scientific cooperation would continue within the International Union of Directors of Zoological Gardens, and passed on his "best wishes for the work of the association as a representative of the interests of all colleagues in the Federal Republic". Zoo Dresden's director Wolfgang Ullrich ended his dutiful account of the Federal Republic's overbearing actions with best wishes to "Ursula and the children". Johannes Schwarz from the animal park in Stendal wrote: "External circumstances prompt me to take this step. I hope you will remember me fondly."[15]

Today, it is difficult to understand the true reason behind these resignations. The mood between the two German states was undoubtedly tense and resentful at that time, with the GDR pushing for international recognition while the Federal Republic continued to insist it was the sole legitimate representative of the German people. The former director of the zoo in Erfurt, Fritz Dietrich Altmann, later recalled that all GDR zoo directors had been summoned to the Ministry for Culture, where they were ordered to leave the VdZ. Dathe was apparently not present,

21 October 1969
Willy Brandt becomes
federal chancellor

7 December 1970
Willy Brandt kneels before the memorial
for the Warsaw Ghetto

Paw prints from young predators are added to the cornerstone of the new Predator House, 1971

but the ministry official said that the matter had been discussed with him.[16] Klös later expressed his certainty that the order had come in response to his election as president of the VdZ at a meeting in the autumn, effective from 1 January 1969. He believed this had displeased Zoo Berlin's competitor in the East.[17] If this is true, it once again highlights the influence that Dathe wielded over the comparatively small zoological community of East Germany, where he was also chairman of the GDR Ministry for Culture's zoo commission.

Lions and more

Following its anniversary year, construction resumed at the Zoo and continued throughout the early 1970s. The former Predator House had remained in ruins since the end of the Second World War, except for the lions' area. Katharina Heinroth and Heinz-Georg Klös had created makeshift enclosures for predators by partially reconstructing former structures. In the late 1960s, attention turned back to construction of a new, modern house for the Zoo's predators that

31 January 1971
Telephone traffic is restored between East and West Berlin

incorporated the latest zoological knowledge. The plans were to retain the attractive savannah-like lion enclosure and add a light-filled building behind it to house the still-growing collection of large and small carnivores. Of course, this new building would also have to meet the stricter requirements now in place regarding housing conditions and keeper safety. Predators are some of the most popular zoo animals with visitors, so the building had to be suitably impressive. The tender was won by architect Paul Schaefers, who designed a three-winged building with a clinker brick façade. The lions' indoor cages formed the centrepiece, with a second large outdoor habitat for tigers added to the building.

In late April 1974, the Zoo was finally able to open its new modern Predator House. In the mid-1970s, "modern" meant that the almost fifty tiled indoor enclosures were easy to clean and could be combined or divided into larger or smaller compartments by means of sliding doors and flaps. The indoor enclosures for smaller predators were lined up along two wings. In the middle of the central building were a small pond and tropical plants. The pyramid-shaped glass roof over this area created a feeling of spaciousness and let in plenty of natural light. The epoxy resin coating on the floor and sophisticated ventilation system were intended to reduce unpleasant odours, meaning visitors to the new house would not be too bothered by the pungent smell given off by the predators.[18] While this may have been the case on the opening day, generations of visitors can attest to the fact that it certainly did not remain that way! Nevertheless, the building was still a milestone in this respect compared to older predator houses.

The opening of the new Predator House was also attended by former director Lutz Heck. Accompanied by Heinz-Georg Klös, Katharina Heinroth, the chairman of the supervisory board and the finance senator, he cut the ribbon outside the new building, which had cost almost 7.5 million marks. While Heck's presence must have been a little awkward for Heinroth, incumbent director Klös saw the invitation as a way of lessening the distance between himself and the former director. He believed that Heck's Nazi past should no longer have any relevance. Special coins and postage stamps were snapped up by visitors on the day, and the requisite performance by the police orchestra was a particular highlight.[19] Berliners also flocked to the Zoo on the days that followed, braving the April weather to queue up for a look at the new animal house.

The new Predator House, 1972

What was once considered an acceptable and even large and attractive facility became an easy target for critics later on. That animal house would end up being completely rebuilt in the not-too-distant future, with many smaller carnivores being sent elsewhere to allow separate habitats to be combined and more species-appropriate conditions to be created for the big cats. Bars and tiles would also be replaced by glass and plants, creating a state-of-the-art Predator House fit for the decades to come.

Love of animals or lack of awareness?

Whereas today's visitors to Zoo Berlin are separated from the animals by ditches or panes of glass, in the past strong bars were more often used. That made it easier for visitors to come into direct contact with the animals, which was not necessarily to their benefit. Immediately after the opening of Berlin's Zoological Garden in 1844, the founding director had to instruct visitors to refrain from handing the animals lit cigars or prodding them with sticks or umbrellas.[20] In 1846, the Zoo urged its visitors to intervene if they witnessed any instances of animal cruelty. In the first six months of its existence, the Zoo lost several of its animals as a result of bad visitor behaviour.[21]

Unfortunately, this problem dogged the Zoo for a long time. In 1947, regular patrols were carried out in co-operation with the Tierschutzbund, Berlin's animal protection association. The volunteers wore armbands and were assigned to different areas of the Zoo by its then director Katharina Heinroth. The primates, big cats and hippos were the most frequent victims of visitor "mischief".[22]

Another major problem since the Zoo's earliest days has been visitors' desire to feed the animals. In the opening year, visitors were prohibited from giving the animals food they had brought onto the premises.[23] But the animal inventory published in the summer of 1845 merely states that guards checked what kind of food the visitors had with them.[24] Many early postcards and photographs show visitors feeding the animals.

Postcard from the Zoological Garden, circa 1900

The feeding of animals – particularly monkeys and great apes – was long considered a highlight of any zoo visit.[25] By the 1930s at the latest, Zoo Berlin had erected signs instructing visitors as to what foods could be offered to what animals and which animals should not be fed at all.

After the war, Berliners were keen to spoil their favourite animals with all sorts of "treats" once again. In the 1950s, Heinroth tried to impose a ban on the feeding of particularly sensitive animals.[26] At the Zoo's 1955 annual general meeting, the board of directors was still against the introduction of a total feeding ban, fearing that would alienate Berlin's "particularly animal-loving" visitors.[27]

That same year, however, an incident served to demonstrate just how dangerous careless public feeding could be when a housewife tipped a shopping bag full of potato peelings into hippo Knautschke's gaping mouth – along with the paring knife. As a first response, Heinroth ordered that Knautschke be fed as much hay as possible in the hopes that this would envelop the knife and allow it to pass safely through his intestines. After three days of nervous waiting, the Zoo and the entire city were relieved to learn that the hippo keepers had found the knife at the bottom of Knautschke's pool. The wooden sections of the handle were completely digested, but Knautschke himself was unharmed.[28] This close call did not serve to bring about any real change, and several animals died in 1959 as a result of visitor "treats".[29]

One year later, female elephant Jambo, who was beloved by the people of Berlin, passed away as a result of incorrect feeding by visitors. This had been a concern for some time, and sometimes her keepers would even paint "Don't feed me" (with non-toxic paint) on her large forehead. But their efforts were in vain. In late March 1960, Jambo fell suddenly and violently ill. Suspicions quickly grew that the illness may have been caused by unsuitable food.[30] While the elephant was being looked after by the Zoo's veterinarian and his colleagues at Freie Universität Berlin, well-wishes poured in from countless Zoo fans all over Germany.

Female African elephant Jambo, circa 1960

Roland IV, circa 1960

Children painted pictures and said they were praying for Jambo; adults suggested treating the elephant with enemas, a special diet, castor oil or alcoholic beverages, or placing her in an inclined position. One person even sent the Zoo twenty laxative tablets.

More than 20,000 Berliners crowded into the Zoo the following weekend to visit Jambo on her "sickbed". When another operation revealed that she was suffering from advanced, incurable peritonitis, Heinroth's successor Heinz-Georg Klös had a difficult decision to make. In the early hours of the morning on 11 April 1960, Jambo was relieved of her suffering. The Zoo was now flooded with a whole new wave of letters. Some expressed gratitude that the animal had been "tenderly cared for" until the very end, while others (usually written anonymously) hurled abuse at Klös – even going so far as to call him a "murderer".[31] On 29 March 1960, the Zoo's supervisory board finally took the necessary step and introduced a general ban on feeding – a policy very much urged by Klös. It seems that people's desire for the sort of direct contact with animals that feeding provides is difficult to manage. The feeders often fail to realise that begging for food is not natural behaviour for these animals.

It is not solely through giving them unsuitable food that Zoo visitors have endangered the health of the animals. When elephant seal Roland IV died in 1961, countless foreign bodies were discovered in his guts during the subsequent autopsy. These were objects thrown into the seal pool that Roland had mistaken for food. For years, a display case containing various items that had been retrieved from the pool stood by the seal habitat as a warning to visitors. Nevertheless, in 1997 another seal died from ingesting foreign bodies. His stomach contained no fewer than sixty-one coins and a pin! Some visitors were clearly confusing the clear water of the seal pool with a wishing well.

Coins found in the stomach of a seal, 1997

A green oasis in the big city: The 1970s and 80s

Creatures of the night: The nocturnal animal house

The completed Predator House featured a central island with greenery, a pond and benches. In 1975, a staircase was installed there leading down to the new nocturnal animal facility that had been created in the building's basement. This dimly lit animal house proved to be a highly popular additional attraction. With its many separate enclosures, it continued Zoo Berlin's aspirations to exhibit the broadest possible spectrum of species within taxonomic families. However, the enclosures were also designed to reflect the animals' habitats in the wild and were arranged according to their geographical origins, which was a clear break with the systematic approach the Zoo had pursued thus far.

Zoo Berlin had long been looking for ways to present nocturnal animals and their dark habitats. In the 1940s, the first successful attempts had been made to show such animals by keeping the lighting dim during the day and making it brighter at night, prompting the animals to retreat to their sleeping caves outside of opening hours. Given the limited amount of space available to the Zoo, it was not able to construct a separate nocturnal animal building. But since windows were not required, the management saw an opportunity in the basement of the newly built Predator House, which had ample space. Under the leadership of the Zoo director's wife and close collaborator, Ursula Klös, the new area was laid out according to geographical criteria. Mrs Klös had herself studied veterinary medicine before moving to Berlin, where she was in charge of the Zoo's animal register.

A tour of the nocturnal animal house leads, therefore, from the landscapes of Africa and Madagascar, through North and South America and on to Asia, before ending in European climes.

Naturally, the creation of a nocturnal animal house also required some new animal acquisitions. The Zoo management wanted the focus to be less on zoological rarities and more on species that were easy to care for while still being interesting for visitors. In their close-to-nature habitats, these creatures should teach visitors about evolutionary similarities that can be found under similar environmental conditions on different continents. Alongside the saucer-eyed slow loris, the first inhabitants of the nocturnal animal

Bengal slow loris Kate, 2019

Slow loris

The slow loris (*Nycticebus*) is a real master of survival. These prosimians are named for their unhurried movements, which help them go undetected by predators. Also, unlike other primates, they do not attract attention with loud calls.

The nocturnal animals live high up in the trees of Southeast Asian jungles, where they spend the day sleeping in hollows. Only under cover of darkness do they set out in search of fruits, leaves, shoots, eggs, insects and even birds, which they creep up on slowly and carefully. Although they have nails rather than claws, slow lorises are excellent climbers. Additional vertebrae give them an especially flexible, bendy spine.

Slow lorises are one of the world's very few venomous mammals. They secrete venom via a gland on their arm, which they lick to make their bite more deadly. They also lick their fur and that of their young to provide additional protection against predators. To "disarm" a slow loris, animal traffickers often remove their teeth. Sadly, that means rescued animals cannot always be returned to the wild.

Slow lorises are endangered. That is partly because their unhurried nature also applies to their reproduction, but the biggest threat comes from hunting and habitat destruction. Slow lorises are sold as pets and used in traditional Asian medicine, as their fur is believed to have wound-healing properties.

These large-eyed primates have been a favourite at the Zoo's nocturnal animal house ever since it opened. Like many other animals who are active at night, slow lorises have a reflective layer behind the retina called a tapetum lucidum (Latin for "bright tapestry") that bounces the little available light that falls on it back through the retina to double the amount of light they have to see by. This layer also makes their eyes shine brightly when caught by a camera's flash. However, flash photography is strictly forbidden in the nocturnal animal house!

Female slow loris at Zoo Berlin, 2019

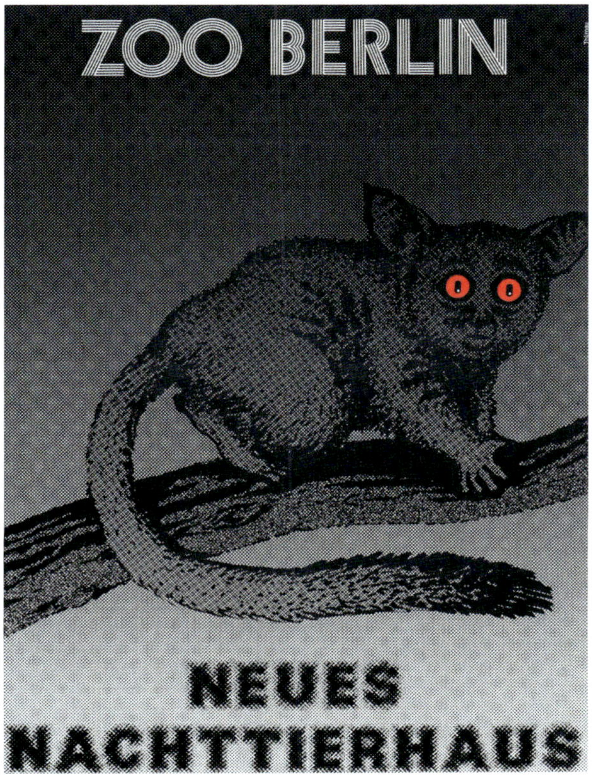

Poster advertising the opening of the new nocturnal animal house, 1975

Kurt Walter, who worked at Zoo Berlin for almost 50 years before retiring in 1990, was the Primate House's heart and soul, shown here circa 1970

house included some of the smallest primates in the lemur family and small marsupials from Australia, as well as armadillos and opossums from South America. There were also desert dwellers like fennec foxes and pale gerbils, bush babies from the African savannah, and bats from all over the world. From the very beginning, the nocturnal house has accommodated two species of bat in a reconstructed "attic", where they impress visitors with their aerial acrobatics.

Underscored by an extremely successful advertising campaign, the nocturnal house's unique appeal quickly made it a popular attraction at the Zoo. The new area, which was completed in four years, was considered state of the art at the time of its opening, much like the new facilities for the elephants, rhinos and apes. The small predators were displayed behind glass, with their enclosures lined up like the tanks in the Aquarium. Inside, individual biotopes had been created using plants and rocks.

Just as important as the new attraction in the basement of the Predator House was the construction of the new Primate House, which began in 1958. As had already happened with the enclosures for feline predators, these renovations implemented the latest zookeeping insights regarding hygiene. The transmission of diseases from visitors had always posed a challenge to the keeping of exotic animals. In 1931/32, almost all the monkeys and apes at Frankfurt Zoo died in a tuberculosis epidemic. Open, difficult-to-clean

1 November 1973
The first two female BVG bus drivers begin scheduled services

16 May 1974
Helmut Schmidt becomes federal chancellor

February 1975
Kidnapping of West Berlin CDU leader Peter Lorenz by anarchist militants

enclosures were identified as one of the main causes of the breakout. Given Zoo Berlin's growing animal stock, the managers knew that living conditions for the animals needed to be further improved. In the mid-1970s the buildings that housed the monkeys and apes were therefore extensively renovated. All indoor enclosures were enlarged by removing the keepers' walkway that ran along the front. The larger enclosures were also filled with far more behavioural enrichment elements – i.e. apparatus for climbing and playing. Large, heavy-duty glass panes were installed to provide better visibility for the visitors and minimise noise for the animals. The easy-to-clean "bathroom-like" interiors of steel, concrete and tiles were retained for reasons of hygiene.

This design was even extended to a completely new addition to the Primate House, the Tropical House, which opened in 1977. Here, visitors can admire other primate species such as mandrills, lemurs and macaques.

The Aquarium is transformed

Aquarium director Werner Schröder retired in 1977 at the age of 70.[1] During his time in office, he had overseen the restoration of the historic building and turned the Aquarium into a respected institution, leaving a solid foundation for his successor to build upon. Schröder had performed important tasks for the Zoo alongside Katharina Heinroth during her tenure, and had played a major role in its recovery. By the late 1970s, at a time when zoos and aquariums primarily measured their status in numbers, his aquarium boasted the largest stock and the greatest species diversity of any public aquarium in the world.

In 1971, Aquarium Berlin displayed virtually all the crocodilians that had been scientifically described at that time, with fifty individual animals.[2] Many of the twenty-four species were represented by only a single pair or an individual. The fact that the size of the crocodile enclosure had not increased since its opening was given less attention than the sheer number and variety of reptiles it displayed. It must have been an impressive sight for people at that time to see so many of these creatures in such a small space.

The Aquarium's new director Jürgen Lange thus inherited a facility that enjoyed a good standing. Nevertheless, the past twenty-five years had taken their toll. The materials available in the post-war period did not have the same durability as modern materials, and various improvements were needed to reflect new developments in aquaristics. What's more, the building's post-war reconstruction had never quite attained the standard of the original. Since renovation was necessary anyway, Lange saw this as a good opportunity to restore the building to its original glory, complete with all its decorative elements and gables – and also to enlarge it.[3] Based on the wealth of experience they had gained during their work at the zoological-botanical garden Wilhelma in Stuttgart, Lange and his colleague Rainer Kaiser designed new ways of presenting the animals in the existing building as well as a new extension to house landscape aquariums. It was clear the project would require funds from the public sector and other donors. The final cost for the renovation of the old building was around three times as high as that for construction of the modern extension. But the entire undertaking was able to be financed with funds from the Berlin state lottery.

The new building opened in December 1980. In addition to a series of large wall aquariums, visitors could marvel at huge tanks arranged

23 April 1976
Opening of the Palace of the Republic

November 1976
Opening of the first women's shelter in West Berlin

16 November 1978
Traffic agreement reached between the two Germanies regarding transit motorways

A spectacled caiman at Aquarium Berlin, 2018

in a hexagon formation that displayed replicas of various biotopes named "Coral sea", "South American river", "African lake" and "Southeast Asia". In addition to fish and amphibians, these tanks also contained the vegetation that would surround their watery habitats in the wild. The layout of these tanks has changed over time. Today there is also a circular tank where various fish, including a small hammerhead shark species, swim around a central rock against an artificial current that simulates their natural habitat. There is also a darkened room dedicated to deep-sea inhabitants. These include pinecone fish with their bioluminescent organs, and chimaeras, or ghost sharks, who usually live in permanent darkness at depths of more than 100 metres. These sorts of underwater worlds help make Aquarium Berlin one of the most visited aquariums in the world. The transition from the old building to the modern extension is barely noticeable in the dimly lit corridors, creating a seamless visitor experience.

In the course of the renovation of the old building, some residents had to be moved or given to other aquariums. Temporary accommodation for the facility's forty crocodiles was hard to come by, so they required a separate solution. The Zoo management responded by converting the interior of the Antelope House into a provisional crocodile enclosure. Each of the scaly reptiles had to be caught, manoeuvred into a box, transported on a truck to the Antelope House, and released into the new enclosure without the use of anaesthesia, which would have posed too

1979
Opening of the Internationales Congress Centrum (ICC) in West Berlin

17 April 1979
First edition of the newspaper *Die Tageszeitung* (*taz*)

great a risk to the animals. The gharials, caimans of various sizes, Nile crocodiles and alligators were first calmed by having a damp towel placed over their eyes. The reptiles were then partially restrained before being placed in their boxes. Given the sheer number of animals in the Crocodile Hall, this task was not without its dangers for the workers involved. Two years later, the entire process had to be repeated in reverse. There were, however, slightly fewer individuals to move, as not all had survived the cramped conditions in the makeshift enclosure. The Zoo had also been forced to give some of its reptiles up for reasons of space, as the veterinarians had noticed an increase in biting and injuries.

The new director continued to gradually reduce the Aquarium's large crocodilian stock. Today, that same area in Aquarium Berlin is home to two gharials, two spectacled caimans, and several turtles. This illustrates the extent to which legislation, public opinion, and above all zoological approaches have changed.

Other animals also had to be relocated during the renovation phases. For example, some fish species were temporarily moved to the basement area. Only the top-floor insectarium was able to continue normally almost without interruption. In 1983, all the work was completed and the new-and-improved Aquarium, with its old and new wing, was presented to the public.

A zoo director with political connections

Those who worked closely with Heinz-Georg Klös made frequent mention of the close connections the director enjoyed with politicians, both in Berlin and at the federal level. He had a reputation for being very adept at gaining financial and/or political support. This trait never drew criticism as it was not for his own benefit, but for that of his zoo. However, there was always a cautionary undertone to this characterisation of Klös. A saying that circulated at the time was that whoever shook Klös's hand would immediately find themselves short a few marks.

It is clear that all zoo directors, wherever they are in the world, have to work constructively with the politicians in their regions, as they often require funding from public coffers. If the land their zoo is situated on is owned by the city, then they are also dependent on municipal services. A zoo attracts tourists; local retailers set up shop in front of its gates; it generates income for the city; it provides a recreational space for local people; and it incurs wastewater and electricity costs. Zoos also provide good publicity for politicians, as the right photo opportunity can promote an individual or convey a message. An environment minister might ask to be photographed with a cuddly polar bear cub, for instance, while other political figures may choose to pose alongside an impressive big cat, or self-deprecatingly present their portly figure next to a hippo. Zoo Berlin animals have even featured on election posters. It is therefore inevitable that anyone managing a zoo will have to interact closely with politicians, particularly at the local level.

Klös generally succeeded in maintaining a good working relationship with politicians of all parties in the Berlin House of Representatives (the state legislature). He was famous among those politicians for negotiating benefits for his zoo. Regular invitations to the Zoo and Aquarium with special tours from the director and his staff helped to maintain those good relations. On such visits, Klös also liked to show his VIP guests new animal arrivals or baby animals that had recently been born at the Zoo.

December 1980
Violent clashes between squatters and police in West Berlin

The politicians showed their appreciation by awarding Klös medals. His wide-ranging efforts were rewarded with the Federal Order of Merit in 1969, the Cross of Merit 1st Class in 1974, and the Grand Cross of Merit in 1982. A star was added to the cross in 1986, and in 1991 he also received the Ernst Reuter Medal, awarded to people who have rendered outstanding services to Berlin. In 2000, Klös was honoured with the title of *Stadtältester* ("City Elder"). The awards were presented by the Berlin mayors Richard von Weizsäcker, Eberhard Diepgen and Walter Momper. It was tradition on such occasions for the famous Schöneberg Boys' Choir to sing. They also sang at his retirement ceremony in August 1991. His wife and colleague Ursula Klös was awarded the Medal of Merit in 1995, as were several other Zoo employees. In 1985 and 1987, the Federal Order of Merit was presented to married couple Elisabeth and Georg Johst, who had been responsible for various departments at the Zoo for many years. Fowl keeper Heinrich Schwarz was one of those who had helped rebuild the Zoo after the war. In 1958, his fifty-seventh year of service, he received the Federal Order of Merit for his work. Katharina Heinroth had already received the Order of Merit in 1957, and was awarded the Order of Merit of Berlin by Mayor Diepgen in 1987. Such honours were also bestowed upon certain members of the supervisory board.

It was standard protocol for the chancellor and president of the Federal Republic of Germany to stop by the Zoo whenever they visited West Berlin from the interim West German capital of Bonn. It was said that Chancellor Konrad Adenauer, who allegedly did not care for Berlin, honoured this tradition only reluctantly. President Theodor Heuss, on the other hand, happily accepted Heinroth's invitation. The aged statesman was driven through the grounds by horse and carriage. Richard von Weizsäcker visited the Zoo several times throughout his political career, first as mayor of Berlin then later as federal president. Chancellors Willy Brandt, Helmut Schmidt and Helmut Kohl also had a special relationship with the Zoo, with Brandt's connection to Zoo Berlin stemming largely from his time as the city's mayor.

But the Zoo's closest friends from the world of politics would be Brandt's successor, Chancellor Helmut Schmidt, and his wife. Hannelore ("Loki") Schmidt came to the Zoo during her

Katarina Heinroth's Cross of Merit 1st Class, awarded by the Federal Republic of Germany, 1957

Willy Brandt with his children at the Zoo, circa 1960

15 June 1982
Opening of the Charité's high-rise hospital building

1 October 1982
Helmut Kohl becomes federal chancellor

Helmut and Loki Schmidt at the Zoo for the handover of the new pandas, 1980. Also in the photograph are Hans Frädrich, Mr and Mrs Klös, and Berlin's mayor Dietrich Stobbe.

husband's first visit to Berlin. Naturally, she was shown around by Heinz-Georg Klös.[4] This was the start of a long-standing relationship between Klös and the Schmidts. Loki Schmidt expressed a great interest in the needs and wishes of Zoo Berlin and knew all about its director's greatest wish: a panda bear. Thanks to her and her husband, this wish would eventually come true in 1980.

In 1983, when the Aquarium was being expanded, Klös wrote a letter to Schmidt's successor Helmut Kohl, who had only been in office for a year, asking if he could procure a pair of Komodo dragons during his upcoming visit to Indonesia. In an attempt to make the prospect more enticing, Klös described the reptiles as "the pandas of the Aquarium".[5] West Berlin's mayor Richard von Weizsäcker added his backing to Klös's entreaty.[6] Sure enough, less than a year later, the Aquarium received a pair of these giant lizards. They had been given to Kohl as a gift from the Indonesian president.

Although it didn't have quite that same level of influence, the Berlin state government wanted to play its part too. In 1984, the president of the House of Representatives, Peter Rebsch, suggested that it was time for Berlin's politicians to help

12 August 1983
Inauguration of the Weltkugelbrunnen
fountain on Hardenbergplatz

the Zoo increase its stock. He said he could make use of his contacts to acquire the "pandas of New Zealand" – kiwi birds – for Berlin.[7]

On 6 September 1986, a pair of kiwis arrived from Auckland. Ever since, chicks of these still endangered birds have regularly hatched at Zoo Berlin. Kiwis are nocturnal, flightless birds that are endemic to New Zealand. The cute and fluffy visitor favourites are roughly the size of a football and sport a beak of up to 20 centimetres in length.

Regardless of party affiliation, the heads of government in Bonn always cared a great deal about what was happening in West Berlin during the Cold War, given its significance for the image and identity of the Federal Republic. However good Klös's negotiating skills were, he would certainly not have been so successful had he been head of a zoo in a West German city like Nuremberg or Hanover, for example. During a tour of Zoo Berlin in 1987, Chancellor Kohl signalled to Klös that he would like to support his animal acquisition wishes because he wanted to do something for Berlin, a city that was close to his heart.[8] The same support was given to Klös's successor from 1991 onwards, Hans Frädrich.

And yet the 1980s were also a decade when the Zoo's relationships with politicians in both Berlin and Bonn were put to the test – starting with a dispute over the Zoo's extension area on the other side of the Landwehr Canal.

The politics of expansion

Since the early 1970s, the management of Zoo Berlin had been planning to build a bridge over the Landwehr Canal leading into the Tiergarten park. This expansion was motivated by a desire to enlarge various enclosures in order to meet

The Zoo's 1981 printed guide

Otto the southern brown kiwi at Zoo Berlin, 2017

25 October 1983
Udo Lindenberg performs in East Berlin

19 November 1983
The one hundred millionth passenger travels through Tegel Airport

27 April 1984
Opening of the Friedrichstadt-Palast revue theatre

The Zoo's expansion plans, with the embassy buildings shaded in red, 1984

23 May 1984
Former Berlin mayor Richard von Weizsäcker becomes federal president

1 October 1984
Opening of the Schauspielhaus concert hall on Berlin's Gendarmenmarkt square

the ever more rigorous, necessary requirements for species-appropriate animal-keeping. If Klös, the avid "collector", was to preserve the quantity and diversity of his animals, the Zoo needed more space. The alternative would have been to give up some animals and thus also his status as head of the world's premier zoo in terms of species diversity. A potential area of land became available around that time: the grounds of the former diplomatic quarter on the southern edge of the Tiergarten. This land was located on Lichtensteinallee, which would form a continuation of the Zoo's central promenade. The site had been badly damaged in the war and had not yet been redeveloped, with only a few partially ruined buildings still standing on it. Using this land to expand the Zoo would fill an empty urban space, green the unused area, and give it an educational function. Klös's plan was therefore met with broad political support. The Zoo envisaged gaining around seven hectares of land, which it intended to use to create landscaped habitats that would house groups of animals of shared geographical origin together. This area of the Zoo would therefore be quite different to the original grounds. A landscape planning competition was held in 1982, and the winning concept was accepted by all sides.

However, this was to be the start of a years-long tug-of-war and political drama full of tough negotiations. Ultimately, the Zoo received not seven, but only three additional hectares. The original concept explicitly included the land where the only two surviving buildings still stood – the remains of the Spanish Consulate and the Royal Danish Embassy. However, some politicians and the Berlin state monument officer raised objections to the demolition of the two structures, stating that the Nazi-era buildings were worthy of architectural preservation. In addition, neither building was the property of the Berlin government, so no firm commitment allowing the Zoo to use the land should have been made in the first place. In fact, the consular office of Spain was still using some of the rooms in the provisionally restored Spanish Consulate building.

In 1983, a media debate broke out over the Zoo's extension plans. Despite some initial major differences, most parties in the House of Representatives soon agreed to approve a development plan for the Zoo that included the site in question. But shortly afterwards, negotiations with the Spanish Consulate stalled. Spain was on the verge of joining the European Economic Community, so it seemed the consulate's office would need the entire building in the not-too-distant future. Then, plans for the use of the site of the former Royal Danish Embassy suddenly emerged, and potential investors came forward. In the press, resistance to the demolition of both buildings was repeatedly linked to the liberal FDP parliamentary group in the Berlin House of Representatives. This junior partner in the coalition under conservative Christian Democrat Mayor Diepgen was campaigning for the buildings to be preserved, and the Christian Democrats were accused of kowtowing to their coalition partner. Diepgen, however, was publicly on the side of the Zoo – as, incidentally, were the Social Democrats. Amidst all these squabbles, time was running out.

The Zoo management did not want to fall out with any of the parliamentary groups in the House of Representatives and was eager to keep the Zoo's expansion plan out of any future election campaigns so as not to jeopardise it. During the planning period, Berlin experienced no fewer than four changes of government. Klös succeeded in involving all parliamentary groups in the project. He also knew it was important not to

11 June 1985
Exchange of 25 agents between East and West on the Glienicker Bridge

5 April 1986
Bomb attack on the West Berlin discotheque La Belle

disappoint the many donors, who ultimately provided a total of almost 20 million marks for the project. In the end, however, despite the director's good political connections and great public support for the plans, the supervisory board had to approve a truncated proposal that would grant the Zoo a much smaller area. Klös had repeatedly pointed out that a reduction in the site's size would severely restrict the zoological plans and that the continued presence of the looming structures would spoil the aesthetics of the complex – but to no avail. This fact was bemoaned publicly in several speeches in June 1984, during the ceremonial laying of the cornerstone for the new site.[9]

Today, the extension area is the only part of the Zoo grounds located on the other side of the Landwehr Canal. It opened on 23 May 1987. For a long time, the area was home to an Africa habitat with South African ostriches, springboks, greater kudus and blesboks, and a South America habitat with guanacos, greater rheas, capybaras and Patagonian maras. With these communal habitats, Zoo Berlin entered new zoological territory. Visitors to the new area could also see emus, parma wallabies and red kangaroos.

The fracas surrounding the creation of the extension area had shown Director Klös his limits and taught him that he couldn't always get what he wanted. However, the Tierpark in East Berlin was still wary of the Zoo's expansion, concerned that its rival could be gaining the upper hand.[10] Ultimately, the expansion was still a success for the Zoo, enabling it to increase its overall size by ten per cent. In the end, history decided the fate of the plot of land that had caused so much controversy. Following German reunification and the relocation of the federal government to Berlin, the former Spanish Consulate became the new Spanish Embassy. The Royal Danish Embassy, meanwhile, was converted into a hotel.

Spies in the Tierpark

In East Germany, the SED regime's power was so extensive that virtually every area of public life was under its control. Those in charge regarded the Tierpark as an instrument of propaganda – a way to present the successes of the Socialist state. The Tierpark attracted over a million visitors each year and enjoyed an excellent reputation around the globe. Its renowned director, Heinrich Dathe, was even permitted to travel to non-Socialist countries. The Tierpark was also a popular destination for foreign visitors and formed part of an international research network. That inevitably meant that the all-seeing, all-knowing Ministry for State Security (known colloquially as the Stasi) had a vested interest in finding out what was going on at the Tierpark. On its northeastern side, the Tierpark bordered the grounds of the Stasi's Lichtenberg district office. In the constantly vigilant and paranoid mindset of an authoritarian regime, it was only logical that enemies of the GDR would use the Tierpark as a vantage point from which to spy on the Stasi premises through binoculars or via listening devices. Special security for the area was therefore deemed necessary.

A glance through the files held by the Stasi Records Agency quickly reveals that an entire network of unofficial collaborators, known as "IMs", was installed at the Tierpark during the 1970s. By the late 1980s, at least eight employees in the Tierpark's middle and upper-level management had been recruited as IMs. Some were headhunted, while others sought contact with the Stasi themselves after believing they had witnessed incidents of concern.[11] Some employees secretly reported on the goings-on at the Tierpark for more than a decade, providing information on the director's trips abroad, for example. Reports

26 April 1986
Chernobyl nuclear disaster

9 January 1987
Opening of the Bode Museum

were also written about conflicts within the management, the Tierpark's economic situation, and the condition of its buildings and structures.

For some time, this network of spies was led by a dedicated case officer, who instructed the individual IMs at secret meetings on the Tierpark grounds or in a nearby apartment.[12] No evidence exists that Director Dathe was ever involved in spying activities for the Stasi. It seems he did not have an overly close relationship with the undercover agents.[13] Dathe loved his profession and would not have done anything to harm the Tierpark. It is clear from Dathe's surviving correspondence with ministries and SED district leaders that there was no reason to doubt his loyalty. Thanks to the network of IMs, the Stasi was able to closely monitor the most crucial moments of his trips abroad, as well as his encounters with colleagues from capitalist countries. The Party's trust in Dathe was so great that the Ministry for Culture even sought his opinion regarding various international organisations and agreements – for example, whether the GDR should join the Convention on International Trade in Endangered Species of Wild Fauna and Flora (CITES).[14]

However, the Stasi also worried about the director's good reputation. As mentioned previously, Dathe had joined the Nazi Party as a student in 1932. He disclosed this former membership in his signed CV, which is contained in one of his Stasi files. In 1981, the Federal Republic published the *Braunbuch DDR,* which contained details of the Nazi pasts of various East German public figures. It included an entry on Dathe. The Stasi's evaluation and control unit promptly called on all departments to check this information and find out whether it had been lied to, whether new information was available, and whether any measures should be taken. The report on Dathe was sent to the responsible district office in Lichtenberg. Apparently, no action was required.[15]

But it seems other authorities also had a need for information. In 1982, East Berlin's mayor Erhard Krack asked members of the Tierpark management to visit the West Berlin zoo on a "spy mission". The Tierpark staff were supposed to provide a detailed overview of the Zoo: What were they planning over there? Should the information gathered serve as a basis for further development plans at the Tierpark? That was, at least, the suspicion of the Tierpark employee and IM who, together with a colleague, was tasked with visiting the Zoo. According to the IM, Zoo Berlin employee Hans Frädrich readily answered all questions asked during the visit from the East.[16] Due to the rather long planning and project development processes that were common in the GDR at that time, it is unclear whether any buildings at the Tierpark directly benefited from this research mission. It is possible that elements of the Crocodile House and Pachyderm House took inspiration from the mission's findings. Dathe got his Pachyderm House in 1988, enabling the Tierpark's elephants to finally move out of the former

Zoo Berlin managers visit the construction site for the Tierpark's new Crocodile House, 1984

horse stable building, which was already in a state of disrepair.

In 1984, Dathe invited his West Berlin colleagues for a visit. Instead of showing them the dilapidated buildings, he led the visitors past the construction site of the Crocodile House and on to the hill formed out of wartime rubble in the northeastern part of the Tierpark. Here, he explained what he intended to build on the site: spacious habitats for mountain animals, with a café at the very top. This must have set off alarm bells for the Stasi, as it would have made their district office visible to a large audience. The IMs reported on these plans, but the Stasi probably assumed that a lack of resources and the bureaucracy involved would be enough to wear Dathe down and perhaps delay the plans until his retirement, which seemed imminent as he had been of retirement age since the early 1980s.[17] Several Tierpark employees, including IMs, were positioning themselves as his successor. The Stasi observed all of this quietly from the sidelines.[18]

Barely any Tierpark informants reported any scandalous or, in the language of the Stasi, "operationally relevant" events. The IMs were therefore often not deemed to be showing sufficient enthusiasm as informants.[19] In fact, sometimes the simplest "offences" were officially reported by the Tierpark itself. In 1984, Dathe complained to the city councillor for culture that entertainers at a children's event had played a song by the West German group Trio – probably the famous "Da Da Da". Another time, a revue performance at the Tierpark festival allegedly made satirical references to the so-called "armed organs", a GDR term for the army and the police force.[20]

Two IMs reported intensively on visitors to the Tierpark from the nearby accommodation for migrant workers from Cuba. The Cubans' alcohol consumption and behaviour were closely monitored. The IMs presumably knew that the SED regime looked unfavourably upon interpersonal relationships between Germans and the "guest workers", as they seem to have been obsessed with interactions between German women and the Cuban men.[21]

In the autumn of 1989, at least one IM informed the Stasi about widespread resentment among Tierpark staff towards the regime. Keepers had apparently read and possibly also distributed leaflets produced by the regime-critical "Kirche von Unten" ("church from below") movement. There had also been a collection of money for political prisoners. The Stasi commissioned three of the IMs to put an end to these activities.[22] Ultimately, however, the only end in sight was for the SED regime and its officers. In November and December 1989, the remaining IMs were deactivated by the Office for National Security, which was the new name of the Ministry for State Security. The undercover agents were no longer needed. Some of those individuals continued to work at the Tierpark.

Celebrations on both sides of the city: Berlin turns 750

Despite the slight setback in the Zoo's expansion plans, it was clear that the grounds would need to look their best for the city's anniversary celebrations in 1987. The extension area on the other side of the canal was to be completed, and the once-prestigious entrance on Budapester Strasse was to be restored. Director Klös predicted that the usually very generous state lottery foundation might be less willing to provide money solely for a magnificent entrance; funds were generally more forthcoming when they were needed for animal houses or operational buildings. It was also

12 June 1987
Ronald Reagan gives his "Tear down this wall" speech at the Brandenburg Gate

30 July 1987
Reopening of the French Cathedral on Gendarmenmarkt square

Pang Pha arrives at the Zoo, 1987

these sorts of structures that had been deemed a priority by the Zoo's supervisory board. Klös therefore included a ticket office and residential building in his plans for the Elephant Gate, as well as a facility for Asian animals directly next to it. This allowed him to "hide" the two million marks required for the impressive gate in the cost of the overall project, which he estimated at nine million Deutschmarks.[23] The lottery foundation was told that the gate and the Asian animal enclosure formed a single unit. The building material for the approved project was supplied by East German state-owned quarry VEB Elbnaturstein. However, the material had to be checked on site, which was no simple undertaking in the Cold War era. As an official business relationship with the company had not yet been established, Klös asked Dresden Zoo for a business trip invitation to enable him to procure the necessary GDR entry visas. His colleagues in Dresden obliged, and so a Zoo Berlin delegation consisting of managers, the architects, and representatives of the transport company and the building materials company set off for Saxony. One of the members only had a Federal Republic passport and no West Berlin identity card, which meant he was not allowed to enter through the inner-city border checkpoint at Sonnenallee. Instead, he had

25 November 1987
The Stasi raids the secret library in the Zionskirche vicarage cellar

to make a detour via Checkpoint Bravo between Zehlendorf in West Berlin and Kleinmachnow in Brandenburg. After a successful meeting and a full-day visit to the plant and quarry, VEB Elbnaturstein was offered the contract. The group only stopped by Dresden Zoo for an hour in the evening to substantiate their alibi. Once the deal was finalised, the director of VEB Elbnaturstein and his employees also had the opportunity to come to West Berlin, including for the opening of the finished gate.[24]

In the city's anniversary year, it was just as important for the Zoo to gain new animals as it was to gain new structures. Klös wanted to use the event as an opportunity to obtain numerous animal gifts, and he was keen to enlist the help of West Berlin politicians and public figures in West Germany. He approached the Senate Chancellery of West Berlin, led by Mayor Diepgen, as well as the Federal Chancellery and the German embassies, requesting that they all ask guests invited to the official 750th birthday celebrations for appropriate animal gifts. Naturally, Klös made mention of the competition in the East in his request. In a letter to Diepgen, he wrote that he had recently learned in Dresden that the Tierpark was expecting animal gifts from other Socialist countries.[25]

The most important step was the creation of a "wish list", which was compiled in December 1986. It included an okapi pair and kagus from France and its overseas territories, a female Indian rhinoceros and red pandas from Nepal, elephants from Thailand, and rhinos from Malaysia and a zoo in the United States. In addition, Klös wanted cranes, pelicans and a Tasmanian devil from Australia, and spectacled bears, anteaters, and vicuñas from Central and South American states. From China, the Zoo hoped for takins, kiangs and deer – but above all, it wanted a replacement for its female panda Tien-Tien, who had died just four years after her arrival.

The West Berlin Senate's protocol department did its utmost to help, compiling a list of the expected guests of honour.[26] In consultation with the Zoo, the mayor then sent largely identical letters to the embassies of the relevant countries in the hopes that all the "birthday wishes" could be fulfilled. Some of the further correspondence was taken over by the Zoo, and the Senate Chancellery kept the Zoo informed of its efforts. Not all the desired animals could be procured, but in May 1987 Mayor Werner Bircher of Berne gifted the brown bear Berni to the Zoo, in the presence of Mayor Diepgen; in October, the young Asian elephant Pang Pha reached the Zoo as a present from the government of Thailand; and on 9 December, Danish environment minister Christian Christensen handed over six new deer. On 2 July of the following year, the Zoo received a gift from the Kingdom of Nepal: the Indian rhinoceros Narayani. She was presented by the Nepalese ambassador Singha Pratap Shah.

Despite the invitation sent to the mayor of Beijing, which the Zoo and the Senate Chancellery had hoped would provide the perfect opportunity, no gifts arrived from China for the anniversary year.[27] The Zoo would not receive new pandas until 2017. Klös's wish for four rhinos from Malaysia also did not come true. But the director of Zoo Berlin was not one to give up easily: if the mayor of West Berlin did not have enough influence, perhaps Federal Chancellor Helmut Kohl could achieve more. Even those efforts were unsuccessful, however. Klös then tried his luck in Indonesia, once again with the support of Chancellor Kohl. But no rhinos could be procured.[28] Nevertheless, with its new inaugurations and animals, the Zoo enjoyed an illustrious year as part of Berlin's 750th birthday celebrations.

11 February 1988
Meeting between Erich Honecker and West Berlin mayor Eberhard Diepgen

March 1988
West Berliners are permitted to stay overnight on visits to East Berlin

It welcomed almost ten per cent more visitors in 1987 than in the previous year.[29]

The city's anniversary was also celebrated in East Berlin – and naturally, the Tierpark played a role. The Central Committee of the Socialist Unity Party even planned a zoo directors' conference at the East Berlin animal park. Dathe was sceptical about this. As the Stasi later learned, he was dissatisfied by how slowly construction was proceeding at the Tierpark and he feared there would be no new buildings for him to show off to his international colleagues, only structures in need of repair. He did not believe that the Pachyderm House, the new Snake Farm, or the new primate facilities would be ready in time.[30] In the end, the Tierpark succeeded in completing the Crocodile House at the very last second and under extreme time pressure. It also completed the renovation of its flagship Alfred Brehm building, where parts of the roof had already collapsed inside the tropical hall. So when Dathe's colleagues from the international zoological world came together at the Tierpark on 13 September 1987, he did have something to show them after all. And according to information provided by a Stasi IM, the visitors were rather impressed.[31]

After all the excitement of the anniversary year, 1988 was relatively quiet. However, one event left fans of Zoo Berlin shocked and saddened. Male hippopotamus Knautschke had been one of the few large animals to survive World War II and had become the progenitor of a long line of hippos born at the Zoo. One morning in June 1988, he failed to mark the enclosure he shared with his son Nante with his excrement, which hippos use to assert their claim to dominance. Nante sensed the opportunity to fight his way to the top of the group and attacked his father.

The keepers were unable to separate the two for several hours. After the fight, an injured Knautschke retreated to a corner of the enclosure, where he became increasingly apathetic. Within days he had lost weight, and eventually he didn't even have the strength to stand up. With a heavy heart, the Zoo management and veterinarians decided to put him to sleep on 20 June 1988. Knautschke was 46 years old. The people of Berlin followed the events closely and were greatly affected by the developments, as the elderly hippo was a public favourite. A hero of countless anecdotes, Knautschke had often featured on the funny pages of Berlin newspapers – with the main subject of the jokes being his weight and his legendary virility. But the star of the Berlin Airlift era was now dead. Many people paid "condolence visits" to the Hippo House, and the Zoo management received hundreds of sympathy letters and telephone calls from all over Germany. With his dramatic life and death, but above all because of his loveable appearance, Knautschke had won the hearts of millions, and they now mourned his loss.

Knautschke and Nante fighting, 1988

19 June 1988
Michael Jackson performs in front of the Reichstag

2 December 1988
Honecker reiterates his rejection of Gorbachev's perestroika reform movement

The animal trade: From colonial take-what-you-want mentality to international breeding programmes

Zoo Berlin's original animal stock was donated from the royal menagerie on nearby Peacock Island. In the early years, the state of the Zoo's finances barely enabled founding director Hinrich Lichtenstein to make any new acquisitions; all he could afford was the occasional small predator, rodent or bird. To procure these, he had to travel to the port cities of France or the Netherlands, where sailors would bring exotic animals home from their voyages around the world. Lichtenstein had to skilfully negotiate a price for the coveted creatures. But over time, the trade in exotic animals became increasingly professional. London-based dealers in particular long dominated the market. Then, in the 1870s, Carl Hagenbeck began building up a trading empire from Hamburg. He went on to be the most important supplier of German zoos for several decades. Other successful animal traders in Germany included the Jamrach family, who had offices in several locations, and the Hanover company Ruhe. Hagenbeck and Ruhe also established their own zoos where they could temporarily "store" their animals before selling them on.

Animal trapper with a Masai giraffe close to Mount Meru in the northern part of present-day Tanzania, 1928

Another important source of stock for zoos was the generosity of well-meaning nobles and other people of influence. The Zoo's annual reports from its first few decades demonstrate the extent to which diplomats, explorers and officials of Germany's nascent colonial empire wanted to make their mark with expensive animal gifts. For a long time, however, zoos had such poor housing conditions that they relied on a constant stream of such gifts or new purchases via animal traders to make up for the high death rate of their animals. Over time, these traders established a dense network of agents in Africa, Asia and South America who procured animals for the growing number of European and North American zoos.

The Europeans and North Americans saw their colonies – especially those in Africa – as lands where they could simply help themselves to whatever they wanted. It was a well-known and accepted fact that for a coveted young animal to be caught, its mother or even entire group would often have to be killed.[32]

A similar take-what-you-want mentality was shown towards the colonies added to the German Empire in the late nineteenth century; the Zoo's annual reports from that time list a number of animal gifts from colonial officials and plantation owners in the new African territories.[33]

Sometimes, zoo directors even went on expeditions themselves to capture animals. On two such trips to East Africa that Lutz Heck undertook in 1925 and 1927/28 he brought back a large group of baboons, five giraffes, three hippos, a rhino, nine zebras, three antelopes, two lions, two cheetahs and thirty monkeys. Usually, these expeditions focused on capturing juveniles. The baboons brought back from Heck's first expedition formed the foundation of the troop that still inhabits the Zoo's Baboon Rock – a visitor highlight to this day. The extent to which Heck viewed Africa as his own personal treasure trove is clear from photographs of his home, where his passion for hunting was apparent throughout. The skin of a maned lioness he had shot himself was spread on the floor. He also proudly displayed two elephant tusks more than a metre in length, a blanket made of twenty-four mantled guereza skins, a lampshade fashioned from the shell of a hawksbill sea turtle, and a stuffed African buffalo. What's more, all his lamps seemed to be made of animal horns.[34] This obsession was not uncommon among zoo directors at the time and clearly demonstrates the prevailing zeitgeist. People back then generally had a very different attitude to the hunting of exotic animals.

Subsequent Zoo directors Katharina Heinroth and Heinz-Georg Klös also undertook or commissioned expeditions to bring back animals, but these were of a different nature. In 1963, Klös and his wife went to South Africa to procure a breeding pair of white rhinos from Umfolozi Game Reserve. These had been "gifted" to the Zoo by the German company Leineweber, which was celebrating its seventy-fifth anniversary.[35]

In 1973, a radical shift occurred for zoos and the animal trading business with the conclusion of the Washington Convention on International Trade in Endangered Species of Wild Fauna and Flora (CITES). This convention was intended to protect all animal and plant species from extinction by eliminating the risk posed by the animal trade. To date, more than 180 countries have signed the CITES agreement, though it is not enforced with equal rigour everywhere. At a CITES conference every two to three years, the parties to the convention and representatives of nature conservation organisations determine the conservation status of endangered species. Some species – including all great apes, great whales, tigers, Asian elephants, grey parrots and most African elephant and rhino populations – are subject to an international trade ban on any captured wild animals. For another group including lions, many bears, all monkeys and dolphins, the exporting countries are obliged to strictly control the trade of specimens.

CITES changed the zoological world by applying legal restrictions to the removal of animals from the wild, a practice that had already faced a good deal of criticism for many years. Over the longer term, this development led to an intensification of large-scale breeding programmes. To coordinate such breeding programmes and the exchange of animals outside their natural habitat, zoos keep so-called studbooks

Transportation of giraffes, circa 1930

Lutz Heck's living quarters at Zoo Berlin, with trophies from his hunting expeditions, circa 1940

that list every animal of a particular species that lives or is born in the affiliated zoos.

Studbooks help preserve the genetic diversity of zoo populations by enabling the exchange of suitable animals between different zoos. The studbooks also contain recommendations for improving conditions. Together, Zoo and Tierpark Berlin are involved in more than a hundred studbooks and European Endangered Species Programmes, making Zoologischer Garten Berlin AG an active participant in more than two-thirds of all international breeding programmes.

One city, two zoos: Zoo and Tierpark Berlin in the 1990s

A visitor boom: Two zoos for all

From the early 1980s, support for the SED regime began declining in the GDR. The economic difficulties and shortage of consumer goods compared to the Federal Republic in the West were becoming more apparent than ever. And in light of the reform movements in Poland and the USSR, the lack of democracy was also becoming increasingly hard to ignore – a fact that was made abundantly clear through the highly visible falsification of ballots during local elections in May 1989. All these factors led to an increase in exit visa applications and a strengthening of the democracy movement under the auspices of church organisations. In the summer of 1989 there was a mass exodus from the GDR via the open borders in Poland and Czechoslovakia, and in the autumn massive nationwide protests took place against the regime. Even a change in the Party and state leadership failed to calm the situation. The new general secretary hoped to ease the pressure by introducing a controlled border opening and new travel regulations. The press conference to announce those changes was held on 9 November 1989 and led to massive crowds storming Berlin's border crossings, bringing about the "fall of the Berlin Wall" and ultimately the end of the SED regime.

Following that historic night, East Berliners and many citizens of the wider GDR flocked to West Berlin, the "showcase of the West", which finally was now easily accessible. Particularly appealing was the bustling area around the "Ku'damm", the famous Kurfürstendamm avenue with its grand KaDeWe department store. On the first weekend after the fall of the Wall, an estimated two million East Germans came to West Berlin. Many East Berliners were keen to visit the zoo "on the other side". For people from the GDR localities of Potsdam, Nauen, Werder, Falkensee, Kremmen and Brandenburg an der Havel to the west of Berlin, Tierpark Berlin had always been difficult to get to, as they first had to travel around West Berlin. Now, they suddenly had a zoo within easy reach. The Zoo management granted visitors from the GDR free admission for the first few weeks following the opening of the Wall, and thereafter charged a greatly reduced admission price. For days, the queue at the Elephant Gate stretched all the way down to Hardenbergplatz. Employees of the Zoo's restaurant sacrificed the

A compass jellyfish, circa 2018

Compass jellyfish "dancing", 2016

Compass jellyfish

The bell of a compass jellyfish (*Chrysaora hysoscella*) has a diameter of up to 30 centimetres. It is adorned with around 15 yellow to reddish-brown radial markings that resemble the face of a compass, hence the animal's name. Approximately twenty long, thick tentacles trail from the edge of the bell. The jellyfish drags these tentacles behind it like a net. When it catches prey, it pulls the tentacles up to its body, transporting the food to its stomach. A compass jellyfish the size of a saucer can eat around 20,000 small crayfish per day. By retracting and extending their tentacles, compass jellyfish are also able to loosen any knots that may have formed. So visitors need not worry if these delicate animals in the jellyfish tank at Aquarium Berlin look hopelessly tangled.

Compass jellyfish have a very broad natural range: they live in the northeastern and southeastern Atlantic from the British Isles to South Africa, as well as in the North Sea and the Mediterranean. These jellyfish are hermaphrodites, usually starting out male before developing both male and female sex organs. This characteristic enables the animal to self-fertilise. Once fertilisation has occurred, the jellyfish becomes entirely female. The eggs develop in the walls of the stomach and are released as planula larvae. The larvae attach themselves to rocks, where they develop into sessile polyps (scyphistomas). Through a process called strobilation, these polyps then segment and separate to form ephyrae, which eventually grow into adult medusae.

As beautiful as these jellyfish are, they can be dangerous for humans. The stinging cells on their tentacles shoot out millions of tiny "poison darts" when they come into contact with prey or an attacker. These also cause a burning sensation on human skin. Treatment with vinegar or urine can help relieve the pain. Incidentally, this stinging mechanism is one of the fastest movements in the animal kingdom.

Visitors queue up at the Lion Gate on Hardenbergplatz, November 1989

tips they had amassed over the course of the year, normally used for their Christmas party, to provide coffee and snacks to the GDR citizens who would start queueing from 6 a.m. in front of the banks around the Ku'damm to pick up their "welcome money". Eastern visitors could also enjoy a bowl of pea stew at the restaurant for a discounted price. During the period of free admission, Zoo employees also distributed free Zoo Berlin printed guides to East Berliners around central locations close to the Wall. Over the two weekends that followed the fall of the Wall, 100,000 GDR citizens visited the Zoo and the Aquarium, which extended their opening hours to accommodate the crowds. Almost four million visitors visited to the Zoo in 1989, just under 900,000 of whom were East Germans.[1]

The Zoo's former director Katharina Heinroth had kept in close contact with her colleagues in East Germany during the Cold War. Unfortunately, she did not live to witness German reunification. Heinroth died on 20 October 1989 at the grand old age of 93.

But the joy and excitement surrounding the fall of the Berlin Wall came with an underlying sense of unease for the Zoo management. They were aware that the almost one million extra visitors in 1989 were a result of the curiosity of East Berliners and East Germans, coupled with the Zoo's advantageous location near the Ku'damm. Now they were worried that many West Berliners, their most loyal visitors, would choose to explore the surrounding countryside rather than visit the "green oasis" in the middle of the Western

February 1989
GDR citizens take refuge in the embassy of the FRG in Prague

2 May 1989
Hungarian border soldiers remove security installations along the Austrian border

city centre.[2] And the number of Zoo visitors did indeed decline over the years that followed, eventually settling at about 3.5 million per year. On the one hand, the fall of the Berlin Wall gave Zoo Berlin the potential to increase visitor numbers due to interest among inhabitants of East Berlin and the surrounding GDR. On the other hand, as a local recreational destination it now had to compete with the green areas outside Berlin and around the River Havel. On top of that, of course, was the competition the Zoo now faced from the Tierpark in East Berlin for visitors old and new – and for public funds.

The transitional period: Fears for the Tierpark's future

Like many East Germans, Heinrich Dathe probably had mixed feelings about the fall of the Berlin Wall. While he would certainly have been happy about the many new freedoms, he must also have felt anxious about what would now become of his life's work. At that time, no German city had successfully hosted two permanent zoos. In fact, Hamburg's first zoo had been forced to close under pressure from competition posed by the Hagenbeck animal park in Stellingen. How would the situation play out in Berlin? Many people were aware that if the city needed to make cuts, GDR institutions would be the first on the chopping block. And indeed, doubts were soon voiced about the viability of two zoos in one city.

At the end of 1989, senior staff at the Tierpark were still working on the final report for the previous year. The first draft had been drawn up in the autumn, in the midst of the flurry and tension of the "Peaceful Revolution" in their country. That document now provides us with something akin to a manifesto of the dying GDR. The numerous subsequent alterations and deletions cast light on the rapidly changing political and social environment, as well as the internal conflicts that emerged as the SED regime neared its end. The foreword drafted by the then 79-year-old Dathe speaks of the effects that the "societal changes in our country" were having on the Tierpark's work. In the edited text, the opening of the Pachyderm House on 29 September 1989 was no longer reported as having taken place as part of the "fortieth anniversary of the founding of our Republic", but was celebrated as a simple, non-political act. Nevertheless, the Tierpark's director could not resist making a subtle jibe: when reporting on visitor numbers, he noted that the Tierpark's grand total of 3.2 million visitors constituted a "foot vote" for his zoo. This comment was in re-

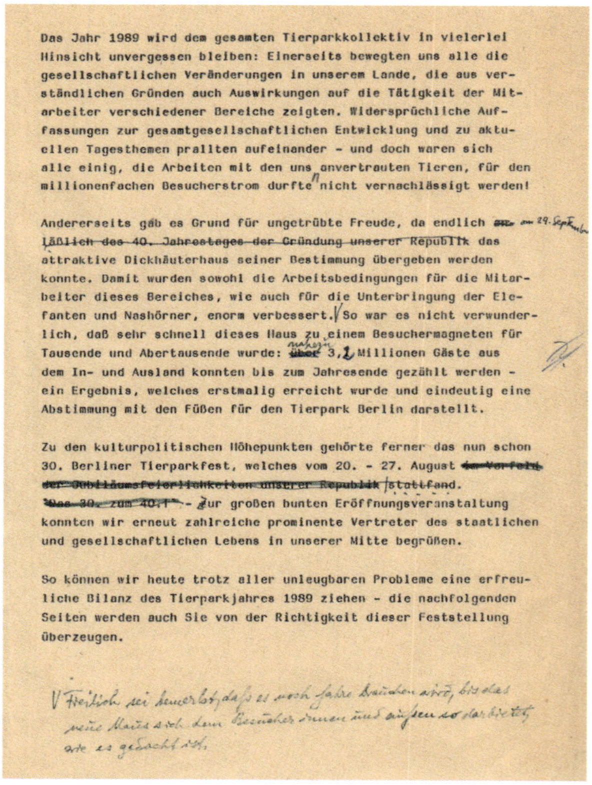

Das Jahr 1989 wird dem gesamten Tierparkkollektiv in vielerlei Hinsicht unvergessen bleiben: Einerseits bewegten uns alle die gesellschaftlichen Veränderungen in unserem Lande, die aus verständlichen Gründen auch Auswirkungen auf die Tätigkeit der Mitarbeiter verschiedener Bereiche zeigten. Widersprüchliche Auffassungen zur gesamtgesellschaftlichen Entwicklung und zu aktuellen Tagesthemen prallten aufeinander – und doch waren sich alle einig, die Arbeiten mit den uns anvertrauten Tieren, für den millionenfachen Besucherstrom durfte n nicht vernachlässigt werden!

Andererseits gab es Grund für ungetrübte Freude, da endlich ~~an-~~ am 29. September ~~läßlich des 40. Jahrestages der Gründung unserer Republik~~ das attraktive Dickhäuterhaus seiner Bestimmung übergeben werden konnte. Damit wurden sowohl die Arbeitsbedingungen für die Mitarbeiter dieses Bereiches, wie auch für die Unterbringung der Elefanten und Nashörner, enorm verbessert. √ So war es nicht verwunderlich, daß sehr schnell dieses Haus zu einem Besuchermagneten für Tausende und Abertausende wurde: ~~über~~ nahezu 3,1 Millionen Gäste aus dem In- und Ausland konnten bis zum Jahresende gezählt werden – ein Ergebnis, welches erstmalig erreicht wurde und eindeutig eine Abstimmung mit den Füßen für den Tierpark Berlin darstellt.

Zu den kulturpolitischen Höhepunkten gehörte ferner das nun schon 30. Berliner Tierparkfest, welches vom 20. – 27. August ~~im Vorfeld der Jubiläumsfeierlichkeiten unserer Republik~~ stattfand. ~~"Das 30. zum 40.!"~~ – Zur großen bunten Eröffnungsveranstaltung konnten wir erneut zahlreiche prominente Vertreter des staatlichen und gesellschaftlichen Lebens in unserer Mitte begrüßen.

So können wir heute trotz aller unleugbaren Probleme eine erfreuliche Bilanz des Tierparkjahres 1989 ziehen – die nachfolgenden Seiten werden auch Sie von der Richtigkeit dieser Feststellung überzeugen.

√ Freilich sei bemerkt, daß es noch Jahre brauchen wird, bis das neue Haus sich dem Besucher innen und außen so darbietet, wie es gedacht ist.

A page from the first draft of the Tierpark's 1989 annual report, with visible deletions and edits

7 May 1989
Large-scale electoral fraud in GDR local elections

June 1989
Beginning of mass exodus of GDR citizens to Hungary

sponse to the many observers who had interpreted the year of mass exodus from the GDR as German people "voting with their feet" in favour of the political and economic model of the Federal Republic in the West. If we take as comparison the number of visitors to West Berlin's Zoo in the previous year (which is the only fair comparison), then it is clear that 1989 was indeed a huge success for the Tierpark. In 1988, Zoo and Aquarium Berlin welcomed about 2.86 million visitors (the same number as the Tierpark).[3] Incidentally, Dathe's jibe did not make it into the final printed version.

Also deleted from the later version of the report was mention of the SED and FDJ (Free German Youth, the SED's youth organisation) secretaries, who previously formed part of the Tierpark's "leadership collective". The list of employees who had received a medal of honour on the occasion of the fortieth anniversary of the GDR was also deleted. However, a list was still published of those who had shown special commitment to the Socialist cause. The activities of the Tierpark's extremely popular youth club were also mentioned, with the exception of the events held in the Young Pioneer camp to mark the fortieth anniversary of the GDR. The draft report proudly announced that participants in the thirtieth edition of the popular Tierpark festival included Politburo member Günter Schabowski, who was also secretary of the Central Committee and first secretary of the SED in Berlin, as well as Central Committee member and mayor of East Berlin Erhard Krack. In the final version, these high-profile names were subsumed under the succinct formula "numerous guests". It was Schabowski who had unintentionally sparked the opening of the Berlin Wall during a press conference the previous month. The edited version completely removed the orchestra of the Ministry of the Interior (which was responsible for the East German police, the paramilitary Combat Groups of the Working Class, and the prison system) from the list of orchestras that performed during the 1989 concert season, while the orchestra of the Felix Dzerzhinsky Guards Regiment (the Stasi's paramilitary wing) was shortened to "orchestra of the guard regiment". In 1988, its full name had been used. In the final version, the interior ministry's orchestra was back on the list. With foresight, editors changed any mention of the "capital of the GDR" to simply "Berlin". In the Tierpark's printed guide of 1989, the term "democratic Berlin" was still used persistently, and it was proudly announced on the inside of the cover that the Tierpark was the bearer of the prestigious Order of Karl Marx and of a Patriotic Order of Merit second class. The following year, there was no mention of those awards. One thus gets the impression that many things were being nervously deleted that, just a short while before, had been deemed appropriate to include, even as a source of pride. At the same time, the latest breeding successes and developments in the Tierpark's animal stock that the report lists clearly highlight the park's zoological achievements.

The managers of the Zoo in the West of the city were also concerned about the future of the Tierpark in the East – or, more precisely, the future of their relationship with the Tierpark. After the fall of the Wall and in preparation for the expected reunification of Germany, discussions about the future of various cultural institutions began all around the country – but especially in formerly divided Berlin, which now had two of many such institutions. Often, as plenty of people had anticipated, it was the East Berlin counterparts that were closed. As a result, many feared the demise of the Tierpark, especially given its dire structural condition in many areas and the large staffing costs. Despite the indefatigable efforts of its director and many of its employees over recent years, things were not going well for the zoo in East Berlin. The shortage economy of the GDR had left its mark. Although the Crocodile House had opened in 1987 and the long-planned Pachyderm House had been inaugurated a little over a month before the fall of the Wall, financing these projects had removed funding from many other areas. The Tierpark's 1989 printed guide reported that the park was "already one of the largest in the world, and can certainly claim to be so after its completion. So far, about 140 hectares have been completed or are in progress."[4] The park was therefore still considered unfinished at that time. It is no doubt thanks to the ingenuity of the Tierpark staff and their passion for their animal park that, even then, Tierpark Berlin was in the top league of international zoos. While structural and technical resources were always limited, zoological successes were plenty. Nevertheless, employees were now highly uncertain as to whether their beloved Tierpark was on the brink of closure.

Coming together

Looking back, we can see that no one ever seriously considered closing the Tierpark in 1990; not the West Berlin politicians, nor the director of the West Berlin Zoo, and certainly not the Tierpark staff and members of the East Berlin Magistrate. It was clear to all involved that not only was the Tierpark a globally recognised zoological institute, its closure would be a political scandal.

The Senate Chancellery led by Governing Mayor Walter Momper of the Social Democratic Party issued clear warnings that no one should give even the slightest impression that the "West Zoo" had any intention of absorbing the Tierpark. The Senate also wanted the Zoo to give first consideration to Tierpark employees lacking job security when making any new hires. At that time, the Senate had no plans to place the Tierpark under the control of Zoo Berlin. According to the Senate Chancellery, all legal issues surrounding a potential future merger would need to be discussed by Berlin's next state parliament, which was soon to be elected. However, the Senate welcomed proposals from the Zoo on the issue.[5] Reading the signs of the times, the Zoo management thought it best to keep their heads down. Given the upcoming elections for the first Berlin-wide government since the Second World War, neither the Social Democrats nor the Christian Democrats, who it seemed likely would soon be the ruling party, would allow uncertainty surrounding the Tierpark's future to pose an electoral risk.

On 21 August 1990, the Zoo and Tierpark signed a cooperation agreement in which they each agreed to "preserve and develop their own special character". Zoo Berlin was to remain a "city-centre zoo with large, impressive animal houses", while Tierpark Berlin was to further develop itself as a "landscaped animal park with spacious, open habitats". But then a process began in Berlin that was occurring in a similar form everywhere else in the soon-to-be new federal states in the territory of the former GDR. The Eastern Magistrate and Western Senate, which began closely cooperating in the spring of 1990, attempted to preserve jobs and business assets in East Berlin and to sustainably reorganise former state-owned companies and cultural institutions. However, this process was usually accompanied by job cuts, downsizing, and the replacement of managers. Comparable companies and institutions in the Federal Republic were used as a yardstick for restructuring in the East.

In 1990, Tierpark Berlin had more than 450 employees – almost twice as many as the Zoo. Many of its buildings, including the prestigious Alfred Brehm building, were in a state of disrepair. The Stasi had demanded reports every winter for several years about the state of the heating systems, whether pipes were freezing and whether animals were dying as a result.[6] Where could things go from here?

Zoo Berlin's management and supervisory boards certainly did not want to take over the Tierpark in its present state. To do so would have meant financial ruin, especially since the Zoo was also reliant on regular state subsidies at that time. But closing the Tierpark was also out of the question. The decision was made, therefore, that the Zoo should at least have control over the direction in which both facilities would develop. After all, any final agreement should not be to the detriment of the older zoo in the West.

At a press conference held in the Alfred Brehm building on 26 October 1990, representatives of the Western Senate and the Eastern Magistrate officially agreed that both zoos should be kept. At a joint meeting on 5 November 1990, the Magistrate and the Senate – now functioning as the "MagiSenat" – decided that a concept for the future of the Tierpark should be presented by the end of the year, developed "in consultation with Zoologischer Garten Berlin". The House of Representatives would then approve this concept after the upcoming elections. The Zoo was asked to develop the concept, present opportunities for cooperation with the Tierpark, and provide statements about the staff and funding that the eastern institution would require going forward.[7] The fate of the Tierpark was thus placed in the hands of its former West Berlin rival.

Similar processes were repeated thousands of times over throughout eastern Germany from the mid-1990s onward. GDR companies all over the country sought advice from the west – or had this advice thrust upon them. In addition to the feelings of resentment that the latter practice in particular must have created, the "lordly manner" of west German reorganisers and consultants was problematic. However, the Zoo had expected that it would be brought into the redevelopment of the Tierpark at some point, and in the autumn the management had already begun to think about what the plans for a shared future might look like. These concepts assumed the viability of two zoos in the city and took into account the emotional attachment that Berliners in both east and west felt towards "their" zoo. They even spoke of a *need* for more than one zoo in order to properly fulfil the new role of zoological institutions and praised the zoological successes of the considerably younger Tierpark. The most crucial factor, however, was the coordination of each facility's thematic orientation, which was meant to ensure that the character of both zoos was distinct and individual while also complementary.

From early on, it was clear that urgent work was required to improve the structural condition of the Tierpark. Of the few animal houses that had been built, many were in a poor state, unlike those at the Zoo in the west. Zoo Berlin management felt that several enclosures simply had to be demolished, including those for the great apes and rhinoceroses, as well as the old cage wagons that were still being used to house the sun bears. And yet the plan was still for both facilities to preserve the diversity of species for which Berlin had become known, with a visit to Zoo and Tierpark allowing animal lovers to experience a wide variety of the world's fauna right here in Berlin. "Must-see" animals that visitors always expect to find at zoos, like elephants, giraffes, zebras,

12 April 1990
Lothar de Maizière is elected prime minister of the GDR

18 May 1990
East and West Germany sign a treaty on economic and monetary union

lions and tigers, were to be kept at both institutions, but the idea was for the Zoo and Tierpark to exhibit different species and subspecies. It was suggested that the Zoo could cease to keep African elephants, for example, and that both institutions should no longer show the same species of bear, giraffe or deer. The Tierpark would keep deer from northern climes, while the Zoo would show tropical species. As for rhinos, the idea was that the Tierpark would focus on white rhinos only. The Tierpark was also to keep all the tortoises. One recommendation that was particularly painful for Tierpark supporters was that all apes should be kept at the western Zoo, given the poor structural conditions at the Tierpark.

The Tierpark also had a very large workforce compared to the Zoo, with almost twice as many employees. Although it was true that the Tierpark covered a much larger area, it housed only about half the number of animals and species as its western counterpart.

In general, the strategic concept developed at that time can be summarised as follows: the western Zoo would boast a city-centre location with numerous walk-through animal houses, presenting particularly demanding or sensitive species in more elaborate facilities and buildings, while the eastern Tierpark was to offer open, expansive facilities that housed large groups or herds. At the Zoo, the organised exhibition of animals according to kinship and taxonomy should be retained, while the Tierpark should focus more strongly on presentation based on geographical origin. The two facilities were to have the same opening hours and entrance fees and be advertised together. Both organisations were to be the responsibility of the same authority – the Senate Department for Finance – to avoid competition for funding. Until then, the Tierpark's funding authority had been the East Berlin culture council.[8] On 23 November 1990, the Zoo's supervisory board decided to conduct preparatory work for a shareholders' agreement for the Tierpark along the same lines as the agreement being drawn up by Berlin utility company BEWAG, which was reuniting at the same time, and help develop a concept that was "reasonable" for the joint-stock company of Zoologischer Garten Berlin AG.[9] As always, the devil was in the detail. One example of this was the planned transfer of all the apes from the Tierpark to the Zoo. While it was true that the ape facilities at the Tierpark were in poor condition, the institution had nevertheless built up a very successful orangutan breeding programme and it may well have seemed as if something valuable was being wilfully destroyed for no good reason.

Unfair dismissal

It is hard to determine the role that the rivalry between Heinrich Dathe and Heinz-Georg Klös played in these deliberations. It seems that Dathe's input was not properly sought or possibly ignored. Given the Tierpark director's advanced age, many expected his imminent retirement. But Dathe feared for his legacy and his staff. He felt that it was his duty to lead the Tierpark through the difficult post-reunification era. Perhaps he was also struggling to relinquish control. After all, the Tierpark had been his main purpose in life for more than forty years. Dathe often emphasised how important it was to him that the Tierpark maintained its independence. On 7 November 1990, Dathe celebrated his eightieth birthday. The media were present at the public celebration, where he was honoured by many of his contemporaries. Once again, Dathe insisted that he would remain at the Tierpark's helm,

17 June 1990
The East German parliament passes an act
to oversee the privatisation of state property

3 October 1990
German
reunification

4 October 1990
The first pan-German parliamentary session
takes place in the Reichstag

Heinrich Dathe with his colleagues Falk Dathe and Joachim Hänsel at the Tierpark's Pachyderm House, 1990

steering it into calmer waters. Many illustrious guests came to visit the director at his Tierpark. Even Zoo Director Klös was present, and gave a speech praising the guest of honour and his life's work. Klös also emphasised: "Not only can Berlin sustain two zoos – Berlin *needs* two zoos!"[10] It is impossible to know whether this statement represented his genuine opinion or was a politically calculated move. It is certainly true that many of his words of praise aligned perfectly with the strategy that had been adopted after countless conversations with politicians.

While many people suspected that Dathe was seriously ill at this point, most were probably unaware of the true state of his health. Shortly after the turn of the year, the Tierpark's founding director passed away.

Just a month before he died, Dathe experienced a severe blow. On 7 December 1990, he was dismissed with effect from 10 December by the Magistrate's deputy city councillor for culture. He was to make way for a younger successor who would lead the Tierpark into a unified future. Dathe and his wife were also to vacate the director's residence, where they had lived with their family since 1955, by the end of the month. That same day, his deputy was appointed acting director. The dismissal was based on a clause relating to the permissible age of GDR public employees that was written into Germany's Unification

2 December 1990
First pan-German
federal elections

2 December 1990
First parliamentary elections in
reunited Berlin

20 June 1991
The Bundestag declares Berlin
the seat of German government

Treaty. It did not come as a complete surprise, as that clause and the pertinent legal situation were well known. Within the Tierpark itself, Dathe's deputies and close colleagues had been trying to position themselves as the boss's successor since 1982 at the latest, when Dathe was already over 70 and therefore seemingly ripe for retirement.[11] It even seems that Dathe was offered the opportunity to step down willingly on his eightieth birthday. However, the founding director was not only deeply attached to his Tierpark, he was also determined to fight for it in a time of widespread rumours about the possible "liquidation" of the institution, and considered himself the best person for the job. He may have also been intending to nominate his own successor. Therefore, the Magistrate's harsh and uncompromising approach must have hit him hard. Dathe's health deteriorated rapidly around the end of the year, and he died on 6 January 1991.

For many people, particularly those in eastern Berlin, the government's actions following the first all-Berlin elections symbolised a typical "arrogance" in western dealings with GDR institutions. Several thousand people attended Dathe's funeral service at the Baumschulenweg crematorium on 17 January 1991. They wanted to express their solidarity with the former director, Tierpark staff and the Tierpark itself, particularly in the face of its possible closure.

The Tierpark gets incorporated

Rumours of the Tierpark's imminent demise continued to circulate throughout 1990. Despite a public statement by the MagiSenat in October 1990 that no such action was planned, the authorities and the Zoo continued to receive a large number of protest letters, and press articles doom-mongered about the closure. Therefore, at the end of the year, the Senate Department for Finance prepared a series of standard responses to questions about the future of the Tierpark and its staff.[12] In April 1991 the Tierpark became the private limited company "Tierpark Berlin-Friedrichsfelde GmbH", wholly owned by the state of Berlin. Over half of the more than 400 employees were kept on.[13]

At the first shareholder meeting, Zoo director Heinz-Georg Klös was elected chair of the Tierpark's supervisory board. Shortly thereafter, his scientific assistant Bernhard Blaszkiewitz, who had been working at Zoo Berlin since 1984, was appointed managing director of Tierpark Berlin-Friedrichsfelde GmbH. Klös retired as director of Zoo Berlin a few months later. At roughly the same time, however, he was elected a member of the supervisory board of Zoologischer Garten Berlin AG.

A lot of resentment was sown during the period of transformation following German reunification. Neither the Zoo management nor the Senate apparently considered the disastrous impression it must have made on the many Tierpark supporters and eastern Berliners for all the new masters to come from the west. Those decisions are difficult to understand today, but they were commonplace across the former GDR and created a great deal of frustration and prejudice. It only slowly dawned on those involved that this approach to the privatisation process was at the very least insensitive. In retrospect it is clear that it was a huge mistake on the part of the politicians in office at that time not to involve the former East Germans more closely. And this mistake would have some very specific consequences at the Tierpark, especially when it came to discussions about the Snake Farm.

30 April 1991
Production of the East German "Trabi" car is discontinued

5 July 1991
The Bundesrat votes in favour of keeping its offices in Bonn

20 September 1991
Start of the first Berlin Wall border guard trials

A venomous atmosphere

LIEBE TIERPARKFREUNDE!

Bitte helfen Sie alle dem Tierpark Berlin-Friedrichsfelde im Kampf um den Erhalt seines Tierbestandes!

Wir als Bürger sind gegen den Beschluß der Aufsichtsräte vom 16.09.1993, daß unsere Reptilien, insbesondere unsere Giftschlangen, in den Zoologischen Garten transportiert werden!

„Tierpark-Ede" nach Erich Schmitt

Dem Zoologischen Garten sein Aquarium, dem Tierpark sein Terrarium!

Für den Bau eines Terrariums mit unseren Giftschlangen als das

"Haus der Gifte"

spenden wir nach dem Motto: Jeder Berliner mindestens 1 DM!

Spendenkonto: "Rettet die Schlangen"
Kto.-Nr.: 940028484
BLZ: 100 500 00
Berliner Sparkasse
Kontakttelefon: 030-892 88 55
Tierparktelefon: 030-51 53 11 31

Spendendruck der ABES-Druckerei GmbH, Rapsweg 53, 12683 Berlin, Tel. 5 62 30 24 / 02, Fax 5 62 30 12

Call for donations to save the Snake Farm, 1993

In October 1993, construction labourers were working hard to restore underground train operations on the U2 line between Wittenbergplatz and Mohrenstrasse stations and thus close the largest gap that the Cold War division of the city had created in Berlin's public transport network. This resulted in an underground line that ran all the way from Pankow to Ruhleben, via major stations including Alexanderplatz and Zoologischer Garten. All such remaining "borders" were removed throughout the early 1990s.

However, a gulf still existed between the Zoo and the Tierpark. They may have reached an agreement to consult with one another and develop a common strategy, but there was still strict separation between the two companies. On 5 October 1993, the Senate agreed that Zoologischer Garten Berlin AG could take over the State of Berlin's shares in the Tierpark. Both zoos were still highly subsidised, so the aim was to make the two facilities more closely linked and thus more cost-effective.

However, many supporters still feared for the future of the Tierpark – and perhaps rightly so, as events in 1993 were to demonstrate. This time the atmosphere was particularly venomous, as the topic up for debate was the fate of the Tierpark's Snake Farm. There had been very few objections to the decision to stop keeping apes at the Tierpark. It was widely regarded as a necessary move given the structural condition of the ape facility, which had always been intended as a temporary solution. In the meantime, the Tierpark's apes had been successfully transferred to various zoos. But the Snake Farm was sacred! Heinrich Dathe had started developing the Tierpark's Snake Farm in 1956 in cooperation with its head Werner Krause. The reptile house had gained a good reputation within a short space of time, especially for the milking of venom from snakes. This is where serums were obtained for use in antivenom for snake bites in large parts of Central Europe. The Snake Farm was therefore a special attraction and a source of pride for many Tierpark supporters. Later, the building was expanded to include a vivarium housing lizards and tortoises.

Unfortunately, the structural condition of the building had not been able to keep up with the facility's success. After reunification, the structural defects and cramped conditions were viewed far more critically by colleagues in the west than by

Poster featuring red pandas that reads "One city – two zoos", circa 1993

the Tierpark staff, who were quite proud of what they had achieved under difficult circumstances and who had become quite accustomed to improvising when it came to the Tierpark's structures.

As early as April 1991, the Aquarium curator published a statement on snake-keeping conditions at the Tierpark that criticised the building's lack of thermal insulation and high energy consumption. He also found fault with the small terrariums, which he said were not appropriate from an animal welfare or species-specific point of view. His safety assessment was also damning, with the visitor corridors deemed unsafe. The curator believed that the building represented a health hazard for keepers and visitors alike, and recommended immediate renovation work. However, he did argue for the preservation of the structure as a shelter for visitors on rainy days, and recommended that the collection focus solely on snakes in order to create more space for the reptiles and provide a more educational set-up.[14] The curator stood by this assessment throughout the period that followed.[15] However, he also expressed his view that, in the spirit of Berlin's zoological institutions becoming more complementary, it would make more sense to show the kinds of animals housed at the Snake Farm exclusively at the Zoo's Aquarium so that the two facilities would not compete against one another. After the initial boom following the fall of the Berlin Wall, the number of visitors to the Zoo had recently dropped again.[16]

The tricky thing about the Aquarium manager's official recommendation was that, ultimately, he was suggesting that the Tierpark's lizards and amphibians be sent to "his" aquarium. However, in a joint meeting on 15 September 1993, the supervisory boards of the Zoo and Tierpark decided that the snakes should be moved to the Aquarium in the west. Tierpark director Bernhard Blaszkiewitz had been consulted at length, but the supervisory board members, all of whom were from the west of Germany or former West Berlin, asked him to leave before the decision was made. The works councils were also not involved in that decision.[17]

For many sceptics, this seemed like another attempt by the "Wessis", as former West Germans were colloquially known, to belittle a GDR institution. The decision also inevitably threatened jobs in that Tierpark department. What's more, some of the crocodilian species offered in

Joint advertising for Zoo and Tierpark Berlin, 1994

exchange by the Zoo were already being kept at the Tierpark.

Along with concerns that the western Zoo Berlin could gain unrestricted access to the Tierpark's animal stocks through the planned purchase of the State of Berlin's shares in Tierpark GmbH, the debate over the Snake Farm caused a great deal of controversy, reviving prejudices and reigniting conflicts.

In October 1993, a petition to save the Snake Farm collected more than 15,000 signatures within a short time and generated a lot of media attention. The conflict not only aroused interest in the regional press, even the national weekly newspaper *Die Zeit* devoted articles to the issue. In the Lichtenberg district council, the PDS parliamentary group, the successor to the SED, suspected an attempt by certain Wessis to dismantle the "former state-owned landmark".[18] The position of the Snake Farm supporters in eastern Berlin prevailed almost everywhere, and the perfectly justified safety concerns could not compete. Faced with such massive resistance, the supervisory board members decided to reverse their decision at an urgently convened meeting on 4 November 1993. The public and media pressure was so great that even local politicians had intervened. The decision was now made to finally

October 1993
Ratification of the Maastricht Treaty by the Federal Republic of Germany

14 November 1993
The Neue Wache becomes the Federal Republic of Germany's central memorial site

elect people from the former East Germany to the Tierpark's supervisory board and to keep the Snake Farm at the Tierpark.[19]

For the time being, therefore, nothing changed for the reptile inhabitants. First, initial renovation measures were carried out to improve the building's safety. For some in the east and west of the city, however, the conflict had helped reinforce the prejudices that already existed about the "other" zoo. And not all of them, it seems, were unjustified.

By 2017, the Snake Farm's structural shortcomings had become even more serious, despite all maintenance efforts. Recurring problems with the roof, dilapidated pipes, mould and pests had become a constant burden. In addition, the terrariums were still too small. Eventually, ticks from the cobras, pythons and mambas even started to plague the keepers. In the end, however, the animals did not go to the Aquarium in the west, but were sent to other zoos or rehoused in other Tierpark buildings.

In the meantime, the Zoo's Aquarium had achieved a small zoological sensation. In autumn 1990, an employee of the University of Wellington in New Zealand contacted Aquarium director Jürgen Lange and offered to lobby the New Zealand government for a few tuataras. The former Berliner had seen the Aquarium during a visit to her old home town and was impressed. She saw an opportunity to develop a reserve population of the extremely endangered lizard-like reptiles through international cooperation. Tuataras are endemic to New Zealand, but are now only found on a few of the smaller islands – roughly 0.1 per cent of their original range. For Lange, this was a unique opportunity to showcase these extremely rare animals. Before that, however, a particularly large terrarium had to be designed for the very sensitive reptiles in order to convince the New Zealand authorities that Berlin had both the experience and the spatial conditions to provide an overseas sanctuary for the animals. Lange has stated it was difficult to find construction companies during the post-reunification period, as all Berlin firms were busy in the east of the city.[20] Eventually, however, the efforts were successful, and the New Zealand government gifted some of the animals to the Aquarium. The sole living representatives of the Rhynchocephalia order, tuataras can live to be a hundred years old and are basically priceless. These were the only individuals of their species in Europe at that time.

New watery habitats for creatures great and small

Amidst these political disputes, the Zoo continued to develop both structurally and pedagogically. The year 1994 marked the 150th anniversary of Zoo Berlin and saw male panda Bao Bao return from his "honeymoon" in London, where he had not fulfilled the hopes placed in him of siring offspring. As there was something to celebrate, it seemed like a good time to bring one of the Zoo's biggest attractions back to Berlin. That year's annual report also emphasised the exceptionally good cooperation the Zoo had experienced with the Tierpark regarding the keeping of individual species and subspecies. Lessons had been learned, it seems.

Meanwhile, improvements were being made in every corner of Zoo Berlin. In the Bird House, small cages were combined to form larger aviaries, and the free-flight hall was improved. A new architectural highlight was finished in the adjacent garden area shortly before the end of the millennium: the Hippo House. This building was opened in September 1997, two and a half years

31 August 1994
Russian forces leave Berlin

8 September 1994
Western Allied forces leave Berlin

after construction began, by Hans Frädrich, who had been Zoo Berlin's director since 1991. Once again, it was the state lottery foundation that contributed a large part of the funds, covering half the construction costs. The Zoo was able to raise the rest itself with the help of donations made by many private individuals and companies. Most of the funds went towards the facility's sophisticated water treatment technology. The massive tank's 650 cubic metres of water has to be constantly cleaned. After coarse materials – mainly hippo manure – are sieved out, fine filtration takes place via several sand filters. This is followed by ozone water treatment and activated carbon treatment. It takes about four hours for all the water in the tank to be cleaned.

Another high cost factor was the glass dome, which spans 1,000 square metres and consists of over 800 individual glass panels. Beneath this impressive roof is the hippos' pool, which has a surface area of 350 square metres and is just over two metres deep at the deepest point. There are also two pools for the pygmy hippopotamuses. To withstand the enormous water pressure inside the tanks, their glass walls are eight centimetres thick. The water in the larger pool is constantly heated to above 17°C, and to about 20°C for the pygmy hippos. Solar panels on the roof

A tuatara, 2019

The Hippo House, 2010

help provide the heating energy, and for much of the year no district heating is needed. At the Hippo House's inauguration, the oldest resident was female hippopotamus Bulette, then 45 years old.

Even more complex than the keeping of hippos, however, is the keeping – and especially the breeding – of jellyfish. These animals are a particular visitor favourite at Aquarium Berlin, despite the fact those same visitors would probably give the slippery creatures a wide berth if they encountered them while swimming in the sea. Floating freely in an aquarium pool, however, these fascinating animals can fully display their magnificent and fragile beauty, causing almost everyone who passes the jellyfish tanks to stop and marvel.

In 1988, Zoo director Heinz-Georg Klös and Aquarium employee Rainer Kaiser visited Monterey Bay Aquarium in California and were intrigued by the jellyfish they saw there. But they knew jellyfish were difficult to keep and impossible to breed. Monterey Bay Aquarium took its jellyfish from the sea and had to replace them regularly. Since transportation of these extremely sensitive creatures is also incredibly complicated, they were practically only ever found in

A Mediterranean jellyfish at Aquarium Berlin, 2019

aquariums near the sea. Although Aquarium Berlin had been the first inland aquarium to succeed in the strobilation of jellyfish polyps in 1932, this knowledge was lost when the keeper responsible for the breakthrough left.

In the 1970s, the aquarium in Düsseldorf succeeded in building a plankton kreisel tank that was able to present the lifecycle of a jellyfish. However, this was a massive, not particularly attractive tank and was not an option for Aquarium Berlin for structural reasons. Nevertheless, when Klös and Kaiser happened upon a few jellyfish during a trip to the Bay of Kiel in 1989, they decided to give jellyfish-keeping another go. Kaiser and his colleagues hooked pipes up to a normal tank in order to generate a flow. With this, the foundations were laid for the successful keeping and breeding of cnidarians. Developers in Monterey subsequently enlarged and advanced the plankton kreisel, but Aquarium Berlin had already made a huge leap towards achieving its goal.

Two special characteristics of jellyfish play a major role in their breeding: their sensitivity to touch and their unique reproductive cycle. Keeping medusae without harming them requires tanks or aquariums to be built with a special flow

Cylindrical jellyfish tank, 2019

of water that allows the creatures to float freely without touching the sides and to move around to feed. Hard or frequent encounters with glass panes or tank walls would lead to injuries. Basically, the pools they are housed in have to recreate the vastness of the ocean. For that reason, all jellyfish tanks create calm water zones in the centre or near the surface of the tank and circulating currents around the sides of the tank so that the jellyfish are swept away when they get too close to the walls. The team at Aquarium Berlin continued to experiment and build new tanks. As a result, great progress has been made since those early attempts, most of which is thanks to longtime jellyfish keeper Daniel Strozynski. Today, Aquarium Berlin has one of the largest jellyfish collections of any inland aquarium in the world, with more than thirty different types. For some time, Berlin was considered the "Mecca of jellyfish breeding".[21] Berlin specialists advise many

Tanks showing the development of jellyfish, 2019

27 October 1998
Gerhard Schröder becomes German chancellor

Algae-breeding tanks at Aquarium Berlin, 2019

zoos across Germany and around the world on keeping and breeding jellyfish. The cylindrical jellyfish tank developed here by former keeper Marco Hasselmann is patent protected.[22]

The breeding of medusae from polyps is particularly complicated. The tiny polyps must be kept at optimum water temperatures and salinity levels, within parameters that vary greatly between the different jellyfish species. Nutrition is also complicated. Some jellyfish live off microscopic algae, which produce sugar compounds in their bodies in combination with light, while others eat small crustaceans. Zoo Berlin breeds its own plankton, which is then used to feed microscopic rotifers, which in turn are fed to the jellyfish. These rotifers (of the genus *Brachionus*) are the smallest creatures at Aquarium Berlin.

Jellyfish are permanently shown in eight display tanks. The predominant species on view to visitors are *Phyllorhiza punctata* (dotted barrel jellyfish), *Aurelia aurita* (moon jellyfish), *Cotylorhiza tuberculata* (Mediterranean jellyfish), *Phacellophora camtschatica* (egg-yolk jellyfish), *Cyanea capillata* (lion's mane jellyfish) and *Chrysaora hysoscella* (compass jellyfish), which is described at the beginning of this chapter.

Modern challenges: The Zoo in the new millennium

Training for creatures of all shapes and sizes

In September 2000, Asian elephant bull Victor arrived at the Zoo from Ramat Gan Safari in Israel. Victor's four-by-two-metre crate was flown on a Lufthansa Boeing 747 from Tel Aviv to Frankfurt, as Berlin could not accommodate aeroplanes of that size. For the pilots of the aircraft, this was a very unusual flight. Normally, cargo does not move, but the nervous young bull's two-and-a-half-tonne weight caused the aircraft to sway repeatedly from side to side. The worried flight captain asked Zoo veterinarian Andreas Ochs, who was accompanying Victor on the trip, if the elephant could be tranquilised. If the animal were to break free, he said, he would not be able to keep the plane on track and it would fall from the sky like a stone. However, Ochs replied that anaesthesia would be too risky for the elephant. Ultimately, the problem was overcome by the great skill of the pilots, who landed the plane in Frankfurt safely despite the difficult conditions. Unfortunately, Victor had to spend a few more hours in his crate in Frankfurt while the veterinarians there checked over the relevant documentation. Then he was loaded into a special vehicle and driven to Berlin, where hopes were high that he would father offspring.

And, indeed, at the relatively young age of ten, Victor mated with Pang Pha, who gave birth to one calf in June 2005 and another in March 2009. Unfortunately, both young elephants died in the spring of 2011. In August 2012, Pang Pha gave birth to another calf, Anchali, who now lives with her mother and aunts Drumbo and Carla in the larger elephant enclosure. Very few zoos keep elephant bulls in the same habitat as the herds of females and their young. What's more, the concept of "protected contact" between elephant bulls and their keepers is increasingly becoming the norm in scientifically managed zoos. Protected contact means that keepers only interact with the bulls via a secure barrier, and there is less and less need to enter their enclosures. However, the elephants – and a large number of other zoo animals – do need to learn certain behaviours and commands. These are taught in special training sessions that condition zoo animals to respond to special instructions with particular actions. This type of training is primarily for medical reasons: it means elephants, for instance, can learn

A group of female sea lions, 2018

California sea lion

The California sea lion (*Zalophus californianus*) is a marine predator. These fast endurance swimmers belong to the eared seal family, which also includes fur seals. Unlike earless or "true" seals, sea lions use their powerful front flippers to propel their streamlined body through the water. When ashore, they can position their hind flippers underneath their bodies and "walk" on all fours. True seals, on the other hand, move by undulating their hindquarters with the help of their well-developed lumbar vertebrae – a mode of getting around that is rather awkward on land. As their name suggests, eared seals also have clearly visible external ear flaps (pinnae), while true or earless seals do not.

The name of the California sea lion is rather misleading, as its range is not limited to California. Outside of mating season, these sea lions can be found all along the western coastline of North America from Mexico to Alaska, where they live in groups. The animals spend most of their life in the water, and are even able to sleep at sea. Their diet consists of fish, crustaceans and squid, which they hunt on long dives up to 100 metres deep. Underwater, the aquatic athletes can reach speeds of up to 30 kilometres per hour. Their whiskers, or vibrissae, can detect even the finest vibrations produced by the movement of their underwater prey.

Sea lions reach around 1.8 to 2.5 metres in length and weigh 100 to 300 kilos, with the males growing considerably larger and heavier than the females. After a gestation period of 11 to 12 months, females give birth to a single pup, which they suckle for up to a year. That means a sea lion mother may find herself suckling a newborn before her one-year-old is weaned.

California sea lions can live to around 17 years of age in the wild and nearly 30 in human care. Zoo Berlin acquired its first sea lion at the 1876 animal exhibition in Antwerp and has kept this pinniped species almost continuously since the early 1900s. Many visitors attend the daily commentated feeding sessions, where keepers share interesting facts about the elegant animals. A large part of the appeal of the feeding sessions is certainly the impressive leaps that the playful sea lions perform to catch their food.

Sea lions underwater, 2018

An elephant show, 1936

to hold their ear or foot out to the barrier for veterinary checks or foot care. But it also allows keepers to prompt the animal to move outside, for example. A positive side effect of the medical training sessions is that they enrich the animals' everyday lives. Training is no longer conducted purely for entertainment purposes, as was customary at the Zoo in the past with the elephants and even the lions. In the 1930s, regular shows featuring thirteen lions took place at the Zoo, led by lion tamer Alfred Kaden.

The training exercises combined with the public feeding sessions are a great addition to the animals' daily routine. They are also a won-

A lion show led by tamer Alfred Kaden, 1938

Spectator seating during a commentated feeding session

derful experience for visitors to Zoo Berlin – especially at the spacious outdoor sea lion enclosure. This and the adjoining Penguin House, built at the same time, form today's "Penguin Hood" complex. The idea of building a new home for the penguins had already been put forward in the early 1990s: although the old one had experienced some breeding successes, it had become too small. The Zoo management and visitors also wanted more spacious pools for the seals. In 1957 and 1959, the dark pool tiles had been swapped for lighter ones so that the animals stood out more clearly against the pool floor and could be better observed underwater. Then, in the early 1970s, an additional pool for grey seals and South African fur seals was created on the northern side. Now, this was to become the site of another major project.

The relocation of the hippos to their new home in 1997 and the demolition of the old hippopotamus facility created room for development in this central area of the Zoo. Two existing pools were combined to create one larger one, while the rear pool was substantially expanded and equipped with a wave machine that has since simulated a "surf zone" for the California sea lions. A large new penguin house, constantly cooled to eight degrees Celsius, was also added, with additional penguin pools as well as pools for otters. The Penguin House is equipped with sophisticated filter systems that allow the Zoo to keep sub-Antarctic king penguins and rockhopper penguins, as these filters are the only way to mimic the almost germ-free natural habitat of these very sensitive animals. Once again, it was the state lottery foundation that generously provided the Zoo with the financial means for this exceptional project.[1]

Since the completion of the seal and penguin complex, visitors can watch swimming sea lions from terraced seating. They can also experience the commentated feeding sessions that take place here every day. Public feedings have always existed in one form or another at the Zoo, but the feeding of sea lions is a particularly good example of the important trend that has now been adopted by modern zoos. During the sessions, keepers introduce visitors to the individual animals, give their names and ages, and share a few anecdotes along with lots of information about the species, their natural habitats, and the threats posed by humans. The aim is to raise awareness, particularly among younger visitors, of the need to protect the ecosystems of numerous species represented in zoos. The keeper may also go into the history of keeping that particular species, or explain certain physical characteristics. The desired end result is for visitors to return home with new and interesting knowledge. Commentated feeding sessions take place with a wide range of animals: the elephants, wolves, wild dogs, chimpanzees, gorillas, king penguins and, of course, the pandas. Audiences find the playful sea lions particularly spellbinding as they speed past the glass walls of their 730-cubic-metre pool and "tease" their keepers. No one minds if they get splashed by the magnificent aquatic mammals plunging exuberantly into the water.

All the sea lions kept at Zoo Berlin can execute learned commands, which is useful for cleaning the pool, for medical purposes, or in the event of danger. The sea lions' medical training sessions also get them accustomed to close contact with humans. The most important purpose of the sessions is to allow veterinarians to conduct health checks, as sea lions cannot easily or safely be put under general anaesthetic, but it also means that the keepers can quickly treat minor problems like fish bones stuck in the throat or small bite wounds. Furthermore, the acrobatic movements and high jumps that the animals are encouraged to perform correspond to how this species would move in the wild and therefore keep the animals physically fit. By now, the sea lions actively demand these behavioural enrichment sessions. But don't be fooled by their playful manner: it would be dangerous for anyone except trained keepers to enter the sea lion habitat – after all, they are still predators with strong jaws and sharp teeth!

The training of sea lions and other animals for the purposes of improving and facilitating their care in zoos is based on a modification of the internationally recognised "operant conditioning" method popularised by psychologist and behaviourist B. F. Skinner. If an animal displays the desired behaviour after a particular command, it is rewarded with a treat. However, punishment for undesired behaviour is never employed at Zoo Berlin. It is more important that desired behaviour be reinforced with an acoustic signal in the instant it is performed. This reinforcement, known as a "bridge", encourages the animal to automatically associate the sound with the action, which facilitates the learning process. A good example of the successful application of this training method at Zoo Berlin is the clicker training technique used with the bonobos. These highly intelligent apes have undergone training for a long time and are now able to perform up to thirty different actions. In their case, the "bridge" is a particular clicking sound that the keepers

Sea lion and keeper in close contact, 2018

Sea lions during a commentated feeding session, 2018

use. Knowing how to physically respond to the signals enables the animals to be examined and treated in a calm environment through bars, and even to be vaccinated without any undue stress. The Elephant bull Victor was also trained in this way from the very beginning – and like many of the animals, he now positively demands his training session every morning. The fact that it is something he *wants* to do makes it much easier for him and the keepers to interact as safely as possible.

Of course, the success of animal training using the classical conditioning method depends on the intelligence of the respective species, but it can also be used to some extent for other animals such as ostriches, giraffes and sun bears. Female sloth bear Kaveri was trained to let her belly be shaved and have ultrasound scans carried out. This was a great help when she was pregnant, as it enabled veterinarians and keepers to separate her from the male at the right time for her own safety. This training method has even been successfully applied in the Aquarium. During the commentated feeding sessions, many visitors are amazed that the gharials Asha and Uma actually understand their names and can be "personally addressed" by the keepers, just like certain domestic animals.

9 November 2004
Guilty verdicts are delivered in the final Berlin Wall border guard trial

10 May 2005
Opening of the Memorial to the Murdered Jews of Europe

King penguins at Zoo Berlin, circa 2010

Knut mania

The polar bear (*Ursus maritimus*) is one of the most popular animals found in zoos. That is partly down to its majestic appearance. Male polar bears can reach up to two and a half metres in length and weigh up to 500 kilograms. The polar bear is the only bear species that feeds almost exclusively on meat. The solitary hunters stalk their prey on or near the ice. Their main food source is ringed and bearded seals, but they will also kill larger animals such as walruses, young beluga whales, or young narwhals. With their highly sensitive noses, polar bears can smell seals through ice one metre thick. Once they have located their prey, they lie in wait at breathing holes or on the edge of the ice.

Polar bears usually mate for the first time at four to five years of age, and the females often give birth to twins. The cubs stay with their mother for up to two and a half years, which is why she only comes in heat every three years. Newborn polar bears are no bigger than a guinea pig, but they grow quickly.

The polar bear's habitat extends over the ice sheets covering the coastal areas of the Arctic – i.e. Alaska, Canada, Greenland, Norway and Russia. Its survival is severely threatened by human

Thomas Dörflein with Knut, 2007

encroachment, human-driven climate change and the extraction of raw materials. Around the world, the polar bear has become a symbol for the need to rethink environmental and climate policy. That is particularly well illustrated by the most famous of all zoo polar bears, Berlin's very own Knut. The adorable little white bear was destined to put Zoo Berlin in the headlines all over the world.

On 5 December 2006 at around 3 p.m., female polar bear Tosca gave birth to two young.[2] Tosca was the last polar bear living in Germany who had been caught in the wild. She was born in Canada and had belonged to the GDR State Circus. The father was a male named Lars, who had been born in Munich and had come to Berlin from Neumünster. Keepers immediately informed the veterinarian of the birth. Regrettably, over the next few hours it became clear that although the young were strong, the mother showed no signs of caring for them. At around 8 p.m., polar bear curator Heiner Klös and Zoo veterinarian André Schüle decided, in consultation with Zoo director Jürgen Lange, to remove the two young animals from their mother's enclosure and place them in an incubator. They hoped that the two cubs could be successfully hand-reared. The newborn bears weighed only a few hundred grams and would have died after just a few hours on the cold floor of their birthing den without their mother's care. Polar bear keeper Thomas Dörflein moved into the cubs' new accommodation so that he could feed the newborns from increasingly larger bottles every two and a half hours. Sadly, one of the cubs died on the fourth day. Now all attention was turned to saving the other cub, who, worryingly, was frequently feverish.

The Arctic is a largely germ-free habitat – and polar bear birthing dens even more so, as they are formed from ice. It is rarely possible to recreate such a sterile environment in zoos. It is therefore particularly important that meticulous attention is paid to hygiene and that external contact with polar bear cubs is largely avoided. That is why the first few weeks are critical for every polar bear born in human care. The Zoo managers initially kept Knut's birth a secret to allow the keepers to get through the difficult first few weeks without intense media attention. They were also probably quite doubtful about the cub's chances of survival.

And yet the little polar bear developed well, and he was able to move into a sleeping box after a little more than forty days. It was now time for the Zoo to announce the happy news of the polar bear birth. It would still be a while, however, before the world was able to see him. In the meantime, Knut began to walk and gain strength. He also got his first teeth, which he occasionally used to play fight with Dörflein.[3]

Until then, Knut's birth was nothing unusual from a zoological point of view. He was simply a cub that had got off to a less than ideal start in life and was now being reared by hand. But for various reasons an unprecedented hype sprang up around the polar bear cub. In the countdown to his first public appearance, local public-service broadcaster Rundfunk Berlin-Brandenburg (RBB), for example, was reporting daily on events at the Zoo.

The plan was to present Knut to the public on 23 March 2007. By this point he would be three and a half months old, the age that polar bear cubs in the wild leave the birthing den and set off with their mother on long walks in search of food. Initially, Knut was to be on view to visitors for two hours a day in the brown bear enclosure (later to become the spectacled bear enclosure), separate from the other polar bears. Knut immediately pulled in the crowds, and the timing of the first public appearance was either a twist of fate or very well chosen. It is true that animal births in zoos have always been special events, and local news outlets often report on them, but coverage of Knut's birth and his unusual start in life were to overshadow anything that had gone before.

Part of the reason for the public's outpouring of love towards the adorable young bear was the intense debate going on at that time about climate change and the threat to the polar bear's natural habitat in the Arctic. What's more, on the weekend following 23 March 2007, a Friday, the heads of state and government of the European Union were meeting in Berlin to celebrate the fiftieth anniversary of the signing of the Treaty of Rome. The focus of the meeting was the ceremonial signing of the Berlin Declaration. This meant that, in addition to the local Berlin media, there were many representatives of the international press in the city. Perhaps feeling uninspired by the dry news they expected to come from the

Knut shortly before his public debut, 2007

summit, they took the opportunity to turn their attention to a cuter, cuddlier topic. And so Knut made headlines all over the world.

The Zoo was in no way prepared for what was to come next. In the days leading up to Knut's "unveiling" and on the big day itself, Zoo employees gave hundreds of interviews. In the beginning, those asking the questions were mainly local media, especially Berlin radio stations. However, on 23 March, almost fifty media organisations came to the Zoo to report on the cub, including international broadcasters like the BBC, CNN, CBS News, and even camera crews from Sweden and Norway, Latin America and Japan.[4] Tens of thousands of additional visitors came over the weekends that followed, keen to catch a glimpse of the young bear. Security guards were hired to usher visitors to Knut's enclosure in staggered intervals during his outdoor hours, and then to usher them onwards so that nobody got crushed and everyone got to see the star attraction. More than 200,000 people visited the Zoo in March 2007, twice as many as during the March of the previous year. The one millionth visitor since Knut's debut appearance walked through the Zoo's gates less than four months later, on 5 July 2007.[5]

Knut and his sponsor, Federal Environment Minister Sigmar Gabriel, at the Zoo, 23 March 2007

To turn the intense interest in the little bear towards a good cause, the Zoo soon created a brand under which it could sell licensed products. Part of the proceeds from the sales went towards projects working to protect the polar bear's natural habitat and to other species protection projects.[6] The first products sold under the label "Respect Habitats. Knut" were the song "Knut geht's gut" ("Knut is doing fine") sung by seven-year-old Lara Trulsen, a book about Knut's first months of life, and jigsaw puzzles with polar bear motifs. Leading politicians also capitalised on "Knut mania". As the official sponsor of the little bear, federal environment minister Sigmar Gabriel drew attention to the problem of human-driven global climate change, a concept which was just starting to become generally accepted. The idea was that Knut should function as an ambassador, drawing attention to the need to protect Arctic habitats. "No polar ice, no polar bear," summarised Gabriel.[7]

This message struck a chord, so much so that Knut even made it onto the cover of fashion magazine *Vanity Fair*. While he had the cover of the German edition all to himself, he shared the cover of the United States May issue with Leonardo Di Caprio, who was promoting his documentary film about climate change *Before the Flood*. The Hollywood actor had been photographed by star photographer Annie Leibovitz, who then came to Zoo Berlin to snap pictures of Knut as well. In spring 2008, a film about the young polar bear and other animals even hit the big screen. *Knut and Friends* told the story of a polar bear family and two brown bear orphans in their respective habitats. In addition, previously unpublished pictures from Knut's "nursery" were

25 March 2007
Celebration of the EU's 50th birthday in Berlin with the 27 heads of state

16 June 2007
Founding conference of the Left Party (Die Linke) in Berlin

30 September 2007
The German national football team defends its FIFA Women's World Cup title without conceding a goal

published.[8] Part of the revenue went to the species protection campaign Respect Habitats.

Meanwhile, a well-known confectionary manufacturer started producing white foam sugar bears in honour of the little celebrity, and teddy bear producers reported an increase in sales of cuddly toy polar bears. The Zoo received a share of the proceeds to help care for the cub.[9]

Knut's bearded keeper Thomas Dörflein soon became almost as famous as the bear in his care. It seems it was the combination of the two that contributed to much of the hype. The unlikely pairing of the clumsy little polar bear with his adorable black eyes and button nose and the sensitive but manly Dörflein proved irresistible for countless fans and the media. When information leaked out that Dörflein played Elvis Presley songs to the bear cub on his guitar in the evenings, a media dream team was born. According to reports from Zoo employees, the rugged keeper was receiving marriage proposals "by the basket-load". Dörflein was not always comfortable with this attention.

The Zoo's director at that time was long-standing Tierpark director Bernhard Blaszkiewitz, who had also assumed directorship of the Zoo in February 2007. Blaszkiewitz took an increasingly critical view of the excitement surrounding the young polar bear and his keeper. Perhaps the conservative zoologist was concerned about the attention being paid to a single animal and, above all, about the anthropomorphism of Knut that was spreading among fans. But Blaszkiewitz was also undoubtedly worried about the safety of his staff. Despite several conversations concerning the risks of maintaining such close contact with the already large predator, Dörflein still entered Knut's enclosure every day. Eventually, Blaszkiewitz put a stop to the "Knut show" and banned Dörflein from playing with the bear.

Knut even received fan mail from the United States, 2008

While this earned him the approval of many zoological colleagues, the Berlin public initially resented the director for this decision.

Dörflein found it difficult to say goodbye to his "adopted son", but he was well aware that it was a necessary step. Many people got the impression that the keeper had very much enjoyed the attention he had received as Knut's sidekick. However, his colleagues described Dörflein as a rather reserved man who was often irritated by his sudden fame. He even refused to accept two television awards for his appearances in TV shows with Knut. On 22 September 2008, Thomas Dörflein died unexpectedly of a heart attack. Countless Berliners laid flowers and letters of sympathy at the Zoo gate and the polar bear enclosure.

Knut himself died on 19 March 2011 at the age of four. He fell into the moat of his enclosure and drowned. Again, the sympathy expressed by Berliners and an international public was over-

24 July 2008
US presidential candidate Barack Obama
speaks at the Victory Column

16 October 2009
Reopening of the Neues Museum
on Museum Island

Knut with his mother Tosca, 2011

whelming. Letters poured in from all over the world in which Knut fans expressed their heartfelt grief. Hundreds of people left goodbye notes for Knut in front of the polar bear enclosure and along the Zoo's perimeter.

Of course, the public desperately wanted to know what had caused the young bear's sudden, harrowing death. It quickly became clear that, although Knut had died from drowning, the cause of his fall was apparently brain disease. It was not until 2015 that a team of scientists from the German Center for Neurodegenerative Diseases, the Leibniz Institute for Zoo and Wildlife Research, and the Charité university hospital established that Knut had suffered from an autoimmune disease.[10] Knut fans set up a memorial stone for the beloved bear at the cemetery in the Spandau district of Berlin where Thomas Dörflein is buried, next to Dörflein's own gravestone. Knut is also immortalised at Berlin's natural history museum as a dermoplastic model (a highly accurate sculpture covered in his pelt), for educational purposes.

Berlin is crazy about bears – a fact that even visitors from outside the city can't help but notice. Berlin's strong association with the animals lends a special importance to the bear enclosures at both the Tierpark and the Zoo. The renovation of the Zoo's bear enclosures built in the second half of the 1960s was therefore subject to careful

June 2010
First postponement of the opening of Berlin Brandenburg airport (BER)

13 August 2011
50th anniversary of the erection of the Berlin Wall

13 July 2014
The German men's national team become FIFA World Cup champions

planning. Back then, the traditional bear pit or cage had been replaced by multiple outdoor enclosures. Over time, however, the demands of the public and zoologists changed, and "quality over quantity" increasingly became the order of the day. Visitors are no longer attracted by the sheer number of specimens and species on display; today's zoo visitors quite rightly want to see habitats that resemble the animals' homes in the wild, where they have plenty of space to move, places in which to hide, and – as far as bears and some other animals are concerned – pools in which to bathe. The reorganisation and restructuring that has been going on almost everywhere for some time has led some zoo enthusiasts to feel that zoos are becoming increasingly homogenous and losing their special, individual character. But the vast majority of zoologists are in agreement with the many other visitors and politicians who are demanding larger and better designed habitats for zoo animals.

After the year 2000, work therefore continued on the conversion of the bear enclosures, under the leadership of directors Jürgen Lange and then Bernhard Blaszkiewitz. This involved merging the black bear and brown bear enclosures to create more space for the brown bears and the neighbouring wolves. In 2012, work began on the complete redesign of the tropical bear enclosure. The old facility had been designed to house a large collection of different bear species and was pioneering in its day. Now, however, all the separating ditches were filled in, the four individual enclosures to the east were merged into two large enclosures, and visitors were given a much more direct view of the animals through glass walls. Some animals were sent elsewhere, including to Tierpark Berlin.

Condolences left on a fence at the Zoo following Knut's death, 2011

During this time, the outdoor facilities for the great apes were also significantly enlarged and made more attractive. The gorilla habitat was extended up to "Lake Lucerne" and the paths around it were made more winding, enticing visitors to set off in search of the primates. For the orangutans, as for the chimpanzees, large net barriers were erected to please both the keen climbers and the curious onlookers. In the first year after the orangutans moved into their new home, a number of holes had to be repaired: the intelligent animals had found weak points in both the netting outdoors and the walls of their indoor habitat, and had been very slowly and patiently working on widening small cracks and gaps.

A Zoo fit for the future

No future without remembrance

For almost the first century of its existence, Zoo Berlin had many Jewish shareholders who were loyal visitors and supporters. That changed from 1933. Those who were able to escape persecution, deportation and murder at the hands of the Nazi regime later reminisced about beautiful summer days spent at the Zoological Garden of Berlin during their childhood. This close relationship between Jewish regulars and the Zoo was described, for example, by Judith Kerr in her world-famous children's novel *When Hitler Stole Pink Rabbit.*

For decades, the Zoo's management never addressed the issue of whether its predecessors might have been responsible for the transfer of Jewish-owned shares below their actual market value. The books published to mark the Zoo's various anniversaries referred to the heavy damage inflicted on the Zoo during the war, but made no mention of the actions of the Zoo management or supervisory board. There was no open dialogue about the Zoo's possible involvement in defrauding Jewish shareholders or discriminating against Jewish supervisory board members and visitors.

At the beginning of the new millennium, the Zoo management was approached by a man named Werner Cohn who wanted to know what had happened to his father's Zoo Berlin shares. Jewish shareholder James Cohn had left Germany for the United States with his family in October 1938 and, like most refugees with Zoo shares, had sold the shares before his departure. Werner Cohn claimed that he had contacted the Zoo back in the 1960s but had never received a response. At first, the Zoo management provided Cohn with information about the buyer, but could not give any further details due to a lack of available documentation. Cohn therefore received no answer to his question about whether the Nazi regime or the Zoo management pressured Jewish shareholders, including his father, to sell their shares. However, based on the assumption that the expropriation had indeed been spearheaded by the Zoo itself, Cohn now asked the Zoo about its moral responsibility towards its former shareholders.

In a move that conveys a sense of alarm and unease, the Zoo management turned to a lawyer

"Meng-Meng", 2017

Giant panda

The giant panda (*Ailuropoda melanoleuca*) is a bear native to central China. With their distinctive black-and-white markings and adorable faces, pandas have long held a special place in many animal lovers' hearts. A giant panda is undoubtedly one of the biggest attractions a zoo can have. Not only does the cute creature have a firm place in popular culture around the globe, it also features as the logo of the World Wide Fund for Nature (WWF) and thus symbolises the threat to its natural habitat.

Like many bears, the giant panda leads a solitary life, only interacting with other pandas during its very brief mating season. Female pandas are only fertile for a few hours or days each year. Pandas generally give birth to just one cub, which reaches sexual maturity at around five years of age. Giant pandas have a life expectancy of around 20 years in the wild, but they can live to the age of 35 in zoos. Given the rate at which humans are continuing to destroy its habitat, the giant panda's slow reproduction rate represents a major challenge to the breeding programmes coordinated by Chinese breeding stations and zoos around the world, which are working to ensure the survival of this threatened species.

"Meng-Meng", 2017

Giant pandas have an entirely vegetarian diet and have evolved to rely on one plant group above all: bamboo. Although pandas can eat other types of vegetation, this sweet grass is by far their favourite. However, bamboo is so low in nutrients that pandas must consume and digest ten to 15 kilos of it a day. As a result, almost all their waking hours are spent sitting upright on the ground, breaking up and chewing bamboo. A regular bear paw would not be able to grasp the long, thick bamboo stems, so pandas have developed a so-called "pseudo-thumb": an elongated carpal bone that is located on the opposite side of the paw to the bear's other digits and that functions like the thumb of a primate.

All pandas living in human care around the world are legally owned by the Chinese government. In the past, the animals were sometimes given as state gifts; today, the government loans the bears to a zoo in or outside China as a gesture of goodwill.

Zoo Berlin got its first giant panda in 1980. In the summer of 2017, the female Meng Meng and the male Jiao Qing arrived and moved into the large, modern Panda Garden in the middle of the Zoo. In summer 2019, Meng Meng gave birth to the first panda cubs ever born in Germany: Meng Xiang and Meng Yuan, better known as Pit and Paule. For four years, they were the new stars of Zoo Berlin. As agreed with the Chinese government, the panda twins left Berlin in late 2023 for the panda breeding and research base in Chengdu in southwestern China, which is where their parents were born. There, they will continue to help ensure the survival of this vulnerable species.

instead of addressing the questions itself. The lawyer's letter, aggressive and dismissive in tone, categorically ruled out that there had been any pressure or coercion in the share sales, while also claiming that anti-Semitic signs had never been put up at the Zoo.[1]

Subsequently, Werner Cohn got in touch with journalists, who also posed justifiable questions about the Zoo's history. Everyone assumed that the Zoo itself had expropriated the shares.

As a result, in 2002/03 Zoo Berlin financed a preliminary study into the sale of shares held by Jewish shareholders prior to and during the Third Reich.[2] In 2011, a commemorative plaque for the shareholders was unveiled at the Antelope House.[3] In 2015, historian Monika Schmidt presented a monograph on the subject, financed by the Zoo. The monograph followed the lives of a number of Jewish Zoo shareholders and provided comprehensive information about the deliberate exclusion of Jewish members of the Zoo's supervisory board. Some of this information had been known since 1994 but had never previously been acknowledged by the Zoo management.[4]

Monika Schmidt's book about the Zoo's Jewish shareholders, 2014

However, those initiatives were not enough for some people with an interest in the debate. They also took issue with the fact that the Zoo was still honouring its former director Lutz Heck with a bust on display in the grounds that featured no comments about his activities during the Nazi era. Today, the bust is in the park opposite the "Empire of Cats" predator house, where the Zoo's former scientific directors from founding director Hinrich Lichtenstein up to Heinz-Georg Klös are immortalised. The bust of Lutz Heck was unveiled at a ceremony in March 1984, one year after his death. The chairman of the supervisory board and the then director both gave speeches, neither of which made any mention of Heck's active involvement in the Nazi regime. Though they must have been aware of at least some of his actions, both men focused solely on Heck's services to the Zoo and his zoological achievements.[5] It seems the time was not yet ripe for a critical reappraisal.

In 2015, following press enquiries and an online petition opposing the public display of the bust, the supervisory board and the Zoo's new director Andreas Knieriem, who had taken office in 2014, decided to properly address the issue. After the author of this book was able to confirm most of the allegations about Lutz Heck in an initial review of the documents that had been catalogued up to that point, the supervisory board commissioned an exhibition on the history of the Zoo for the foyer of the Antelope House.

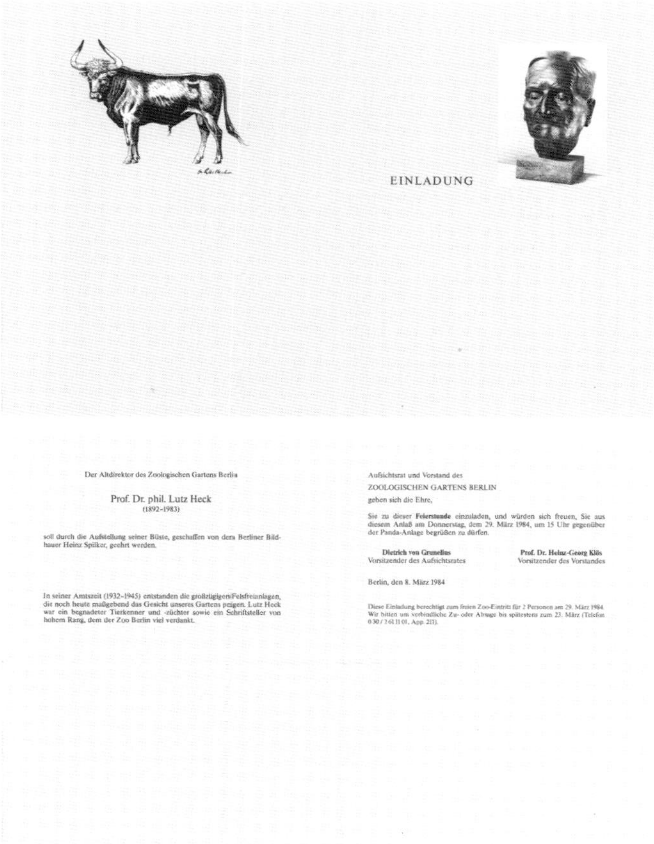
EINLADUNG

Der Altdirektor des Zoologischen Gartens Berlin

Prof. Dr. phil. Lutz Heck
(1892–1983)

soll durch die Aufstellung seiner Büste, geschaffen von dem Berliner Bildhauer Heinz Spilker, geehrt werden.

In seiner Amtszeit (1932–1945) entstanden die großzügigen Felsfreianlagen, die noch heute maßgebend das Gesicht unseres Gartens prägen. Lutz Heck war ein begnadeter Tierkenner und -züchter sowie ein Schriftsteller von hohem Rang, dem der Zoo Berlin viel verdankt.

Aufsichtsrat und Vorstand des
ZOOLOGISCHEN GARTENS BERLIN
geben sich die Ehre,

Sie zu dieser **Feierstunde** einzuladen, und würden sich freuen, Sie aus diesem Anlaß am Donnerstag, dem 29. März 1984, um 15 Uhr gegenüber der Panda-Anlage begrüßen zu dürfen.

Dietrich von Grunelius — Vorsitzender des Aufsichtsrates
Prof. Dr. Heinz-Georg Klös — Vorsitzender des Vorstandes

Berlin, den 8. März 1984

Diese Einladung berechtigt zum freien Zoo-Eintritt für 2 Personen am 29. März 1984. Wir bitten um verbindliche Zu- oder Absage bis spätestens zum 23. März (Telefon 0 30 / 2 61 11 01, App. 211).

Invitation to the inauguration of the bust of Lutz Heck, 1984

The curator of that exhibition and author of this book was granted unrestricted access to all the documents in the Zoo archive and his work was overseen by an external advisory council. The supervisory board and the director agreed that the only appropriate response to the allegations was complete transparency.

The permanent exhibition, which opened in December 2016, is dedicated to the entire history of the Zoo, with a special focus on the Third Reich. All known materials up to the time of the exhibition's opening were analysed, and many previously unpublished documents were included. This book expands on the research findings presented in that exhibition.

In a further bid to ensure full transparency, a plaque was added to Heck's bust summarising his "National Socialist career". The management decided against removing the bust, which some parties had requested, as it did not want the Zoo to be seen as attempting to gloss over that chapter in its history.

In 2015, the Zoo Berlin supervisory board decided to further address its historical responsibility by supporting a special study programme. The Ludwig Armbruster Fellowship Program of the Freie Universität Berlin and the Hebrew University of Jerusalem offered PhD students in the fields of veterinary medicine, biology, ethics or history at the university in Jerusalem the opportunity to work and conduct research for three to twelve months at the FU in Berlin. The scholarships were financed by Zoologischer Garten Berlin AG. The aim of the programme was to strengthen scientific exchange between Israel and Germany. The first grant recipients were welcomed in October 2016, and the project ended in 2023.[6]

Panda diplomacy: Precious bears for Berlin

Around the world, the exchange of gifts is a common way to honour and celebrate friendship. Animals – especially "national animals" – have long served as ambassadors for their respective country. That was certainly the case, for instance, with the bald eagle Willy Brandt who came to Zoo Berlin from the United States. The concept of "panda diplomacy" should be understood against that backdrop.

At the end of the Cultural Revolution in the People's Republic of China, the country was keen to improve its foreign policy relations. From the early 1970s, Britain, France, Spain, Mexico and Japan all received pandas as state gifts from the Chinese government. But China was also eager to develop a closer friendship with the Federal Republic of Germany. Federal Chancellor Helmut Schmidt, who took office in 1974, shared the desire to foster closer German-Chinese relations. In 1975, he became the first German chancellor to visit the People's Republic of China, where he met with head of state Mao Zedong. It is likely that the Zoo's director at that time, Heinz-Georg Klös, was hopeful even then that the trip might result in the coveted gift for his zoo. However, Chancellor Schmidt did not fly back to Germany with a panda in tow.

A few years later, in 1979, Mao's successor Hua Guofeng made a return visit to the Federal Republic of Germany. As previously mentioned, Klös maintained a close relationship with Chancellor Schmidt and his wife. It is said that Loki Schmidt beseeched her husband to raise the "panda question" during Chairman Hua's visit. Chancellor Schmidt apparently later told Klös that Loki even wrote "Panda" in large letters on the bathroom mirror with toothpaste to ensure he would not forget.[7] Ultimately, those efforts were not in vain.

Following the state visit, the Zoo management learned from the Chancellery that it should start preparing for the possible arrival of giant pandas. The Zoo immediately began drawing up plans for a large, glass-panelled enclosure close to the Predator House. However, the potential future occupants remained a secret. It was not until the summer of 1980 that a government spokesperson announced that a pair of giant pandas were expected to be coming to Berlin as an official state gift from the People's Republic of China.[8] On 5 November 1980, Bao Bao ("treasure") and Tjen Tjen ("little heaven") – both of whom had been caught in the wild – arrived in West Berlin. The pandas settled in well and became national stars overnight, with few able to resist the bears' adorable faces. The pandas' arrival resulted in a considerable increase in visitor numbers, as can be seen in the Zoo's annual reports from 1980, 1981 and 1982. Unfortunately, however, hopes of panda babies remained unfulfilled: Tjen Tjen died in February 1984 following a short but severe viral infection, before Bao Bao reached sexual maturity.

Hundreds of animal lovers flocked to the panda enclosure to express their condolences. The

Bao Bao and Tjen Tjen, 1981

Panda enclosure, 1990

Arrival of the pandas at Zoo Berlin, 2017

Chinese press also reported on the bear's death, and the tabloids hyped the event up to such an extent that one could have been forgiven for thinking nothing this significant had happened since the assassination of John F. Kennedy. Rumours spread that the Soviet secret police had poisoned the panda in an attempt to damage German-Chinese relations.[9] There would be a similar response to the death of young polar bear celebrity Knut more than two decades later.

Immediately following Tjen Tjen's death, calls were made for a new mate to be found for Bao Bao. Without delay, the Zoo management sought political support from the Senate Chancellery, the Federal Chancellery, and the Office of the Federal President. The first two institutions passed the request on to China. However, the Chinese government initially rejected this appeal. Pandas were becoming increasingly important symbols, especially as they were an endangered species. Another attempt, once again in vain, was made in 1987 by the Zoo, Chancellor Helmut Kohl, Mayor Eberhard Diepgen and the German embassy in the People's Republic of China.[10]

Finally, in April 1995, Zoo Berlin received a female panda from the zoo in its twin city Beijing. This panda was leased rather than gifted, however, as in the 1990s the Chinese government switched to a loaning system for its panda diplomacy activities.[11] Yan Yan ("the beautiful one") did

Ceremonial opening of the Panda Garden in the presence of Chancellor Angela Merkel, President Xi Jinping and his wife, and Zoo Director Andreas Knieriem, 5 July 2017

not conceive any offspring with Bao Bao, although Zoo Berlin extended her stay and conducted several attempts at artificial insemination. In March 2007, Yan Yan died, leaving Bao Bao the sole panda in Berlin once more. That summer, two chairmen of the supervisory board of Zoo Berlin promptly turned to Federal Chancellor Merkel with the now almost routine request for mediation, initially in vain. Bao Bao died in late August 2012. At around 34 years of age, he was probably the oldest panda living in human care outside China.[12] Berlin now found itself "pandaless" for the first time in over thirty years.

But that would not remain the case for long. Zoo Berlin had good ties to the Chinese authorities and diplomatic relations between the two countries were positive, so it was largely expected that the Zoo would be granted new pandas before too long – and yet the speed at which that happened was quite surprising. Again, the Senate and the Chancellery played their part. Chancellor Merkel highlighted the issue on her visit to China in 2015, and she received positive signals even then. In October 2016, the news came that the Chinese government was officially planning the next panda loan to Berlin.[13]

The Zoo was now under time pressure. To emphasise the significance of this generous gesture, a handover ceremony akin to a state occasion was planned. It was decided that the event should take place in July 2017 to coincide with the G20 Summit – the meeting of heads of state or government of the world's twenty leading industrialised and emerging economies – in which both the Chinese premier and the German chancellor would be participating. Uncharacteristically for Berlin in this day and age, the Zoo was ready in good time! The topping-out ceremony for the new panda habitat took place on 21 February 2017, in the presence of Berlin's Governing Mayor Michael Müller and Chinese Ambassador Shi Mingde.[14] The Panda Garden, declared by experts to be the most beautiful panda habitat in Europe, was completed in just fourteen months. Amidst great media attention, the bears landed in Berlin in late June 2017. They were then given time to get used to their new surroundings out of the public eye.

Finally, the big day arrived: on 5 July 2017 President Xi Jinping and Chancellor Merkel jointly opened the Panda Garden. "Berliners have a very special connection to and a particular fondness for bears," said Merkel, while Xi expressed his hope that the "two very lovable diplomats" would be "special ambassadors for our two countries", emphasising that the event was a symbol of the friendship between Germany and China.

As Zoo Berlin already had plenty of panda-keeping experience, the two bears were loaned to the Zoo for a generous fifteen years. In return, the Zoo pays an annual loan fee that is used to finance in situ species conservation efforts in China. The cute and cuddly appearance of giant pandas makes them one of the most popular zoo animals around the world. These visitor magnets are therefore ideal ambassadors for their homeland and natural mountain-forest habitat. The giant panda is also a "flagship species" as it represents an entire ecosystem and the threats that ecosystem faces. An unforgettable encounter with a panda makes zoo visitors more aware of the importance of species conservation and can awaken a desire to protect endangered animals. Keeping pandas therefore makes logical sense for zoological gardens outside China that are looking to fulfil their educational mission and continue to function as advocates for nature in the twenty-first century.

Standing on its own two feet

The tremendous impact of animal ambassadors became especially clear in 2007 with Knut the polar bear. Although the media hype surrounding the young bear has been judged critically in retrospect, at the time the increased attention that Knut attracted made many people more aware of the shrinking ice caps in the polar regions. For Zoo Berlin, the hype had even further-reaching beneficial side effects. Thanks to the high visitor numbers resulting from the polar bear's birth and the accompanying sale of merchandise, the Zoo was able to generate considerable additional income and thus increase its reserves, creating a good, solid basis for renewed financial independence.

Until the early 1950s, the Zoo had always been largely self-sufficient as the oldest Prussian joint-stock company still in operation. Of course, it had received the occasional government grant, but on the whole it was able to finance itself through ticket sales, lease income and investments. Historical events then resulted in a prolonged phase of dependence on subsidies. All of Berlin's state governments have been very generous in their support for the Zoo and, alongside the Berlin state lottery, have always provided urgently needed funding. During the Cold War, this was partly because Zoo Berlin functioned as

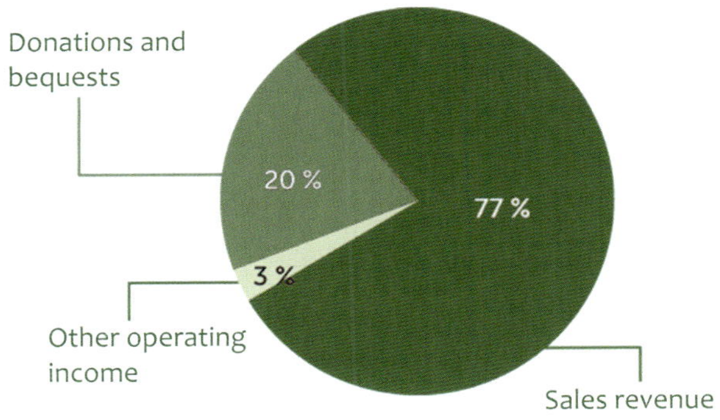

Proportion of donations and bequests in the Zoo's income, 2018

an important showpiece for West Berlin. All Zoo Berlin directors have been extremely grateful for this support, and for the fact that politicians have always stood by "their" Zoo.

And yet it is true that the provision of financial support comes with certain obligations and restrictions. In 2012, Zoologischer Garten Berlin AG was therefore proud to announce that it would no longer be receiving regular funding from the state of Berlin. Since then, the Zoo has been financially independent once more. This relieves pressure on the budget and makes both the Zoo management and the shareholders happy.

The loyalty of Berlin's citizens is just as important as that of its politicians. The Zoological Garden began as an educational project funded by the Prussian king and evolved into a successful non-profit joint-stock company steered by committed members of the city's bourgeoisie. Today, the Zoo still relies on its loyal visitors and the citizens of Berlin. Besides ticket sales, its existence is largely ensured by donations and bequests that clearly demonstrate the Berliners' love of animals.

The loyalty shown by Berliners is not limited to the Zoo and its Aquarium; it also applies to the Tierpark. Since 1993, Berlin's three large zoological institutions – the Zoo, Tierpark and Aquarium – have been grouped under a single umbrella brand. Each of the three facilities is unique, with its own strengths and focuses. Zoo, Aquarium and Tierpark Berlin are valuable recreational facilities that also conduct important educational work and are committed to species conservation and research. All three institutions fulfil the main task of a zoological facility: promoting a closeness to and love of animals.

Making space for big plans

In 2014, Andreas Knieriem assumed the position of CEO of Zoologischer Garten Berlin AG and managing director of Tierpark Berlin-Friedrichsfelde GmbH. Changes quickly became visible; even visitors could tell that a new wind was blowing. One immediate measure initiated by Knieriem was the renovation of the bird house, which had only just been completed in 2013. Shortly after the opening, deficiencies became apparent in the ventilation and heating systems, and the building was also starkly utilitarian in design. As part of the renovation project, the two rather small free-flight halls for African and Australian birds were combined to form a single African free-flight hall of around 450 square metres in size. A separate free-flight hall was created for the birds of Asia. Dense vegetation, natural flooring and rocks made of natural and artificial stone allow visitors today to gain an impression of the inhabitants' homes in the wild and encourage the birds to be active and exhibit natural behaviours. Large windows in the roof let in plenty of natural daylight. To prevent the birds from flying into the glass, the windows are coated with a film that refracts ultraviolet radiation. This coating is visible to birds but not to humans. In the World of

Birds – as the house has been called since its re-opening – geographical division was introduced and the birds from Africa, Australia, Southeast Asia and South America are now presented in four aviary areas known as "loops".

But the most important, and initially the largest, new construction project was the Panda Garden. This animal house is located at the heart of the Zoo and covers over 5,000 square metres. It consists of a Chinese-style pavilion, two indoor areas, and extensive outdoor spaces with climbing and bathing facilities. The indoor habitats have panoramic windows that separate them from the visitor area, where an exhibition provides information about pandas.

Another construction project that resulted in a new visitor highlight was the redesign of the former Eagle Rock. In 2018, after three years of building work, the new Eagle Canyon opened. Once they have passed through a double set of gates, visitors find themselves on a winding path lined with vegetation. The "steppe" section of the habitat features a waterfall and a small stream and is home to vultures and other raptors. This landscape is followed by a "Nordic forest", where snowy owls and great grey owls can be observed up close. In the course of the urgently needed renovation of the historic rock feature, multiple aviaries were merged to create two spacious walk-through facilities. The feathered inhabitants now live in a naturally designed environment alongside other birds from the same region.

Hyacinth macaw in the World of Birds, 2019

Berlin bears

If there is one kind of animal in the German capital that deserves special mention, it is undoubtedly bears. The first successful rearing of a polar bear cub after Knut caused much excitement throughout Berlin and beyond. When Berlin's then first-division Bundesliga team Hertha BSC submitted a video application to become the cub's sponsor, it seemed like a perfect match. The female cub was quickly christened "Hertha", and the Tierpark benefitted from visibility at the football club's home ground, the Olympic Stadium. Posters of Hertha were plastered all over the city, and all eyes were on the young bear when she finally made her debut public appearance at Tierpark Berlin on 2 April 2019. In no time, little Hertha became a visitor favourite, capturing hearts all over Germany. Only one other piece of zoological news drew more media attention that year: the birth of Germany's first giant panda cubs at Zoo Berlin, on 31 August 2019. This highlight of the "Berlin Year of Bears" was accompanied by almost daily international media coverage. The year 2019 was the first in the history of Berlin's zoological gardens where the Zoo, Aquarium and Tierpark registered a total of 5.5 million visitors. This visitor record firmly cemented Berlin's zoos' place alongside other top-ranking zoological facilities in Singapore, San Diego and New York, for example.

Great grey owl in Eagle Canyon

A topsy-turvy world

Following that historic year, in spring 2020 the world suddenly turned upside down – including for Berlin's zoos, which remained closed for six long weeks as the Covid-19 pandemic gripped Europe. Celebrity panda brothers Pit and Paule enjoyed their long-awaited first outing with no adoring public, and polar bear teen Hertha played in front of empty stands. Not only did the pandemic put a sudden stop to the year-on-year increase in overall visitor numbers, but Zoo and Tierpark Berlin also experienced something of a role reversal. At Zoo Berlin, which was usually a tourist magnet in the heart of the busy City West area, a ghostly emptiness hung over the facility even after it reopened on 28 April. But out in the east, the Tierpark, which had always come second to the smaller but far more central Zoo in terms of visitors, was less impacted. While short journey times and a high number of tourists in the western city centre had previously been an advantage for the Zoo, the strengths of Europe's largest animal park became especially apparent during the pandemic: extensive grounds and a high number of loyal regulars committed to the beloved attraction in good times and in bad.

With social life largely suspended, 2020 ended up being one of the most difficult and financially challenging years since the Second World War for Zoologischer Garten Berlin AG and its subsidiary Tierpark Berlin-Friedrichsfelde GmbH. Throughout the six weeks from mid-March to late April that the Zoo and Tierpark remained closed – and the almost six months of closure for the entirely indoor Aquarium – the animals still had to be cared for, the gardens and buildings, some of which are listed structures, had to be maintained, and construction projects had to be continued. Then, in early 2022, the war in Ukraine made the outlook even bleaker, and in Berlin – a city already known for its gruff attitude – optimism became a scarce resource. But thanks to the creativity and drive with which Zoo and Tierpark employees met all the new challenges, the years of crisis were successfully weathered.

You reap what you sow

The foundations for the Zoo's sustainable development have been laid gradually over many years. After an application for EU Regional Development funds, 2018 brought the good news that Zoo and Tierpark Berlin were being awarded € 100 million in GRW funding over several years. Obtaining this level of funding is one task; allocating the money properly and responsibly is another. Getting to grips with funding application processes, EU law and other building policy issues required lots of personnel and time, particularly in the area of technology. But those who sow timely and carefully are sure to reap the rewards, and the results of all that hard work became apparent in the years that followed – mainly at the Tierpark, but also at the Zoo. The opening of the Panda Garden in 2017 was followed by the renovation of the Predator House. The redesign of the animal house from the 1970s included the addition of natural ground materials, climbing trees and bathing ponds, bringing the building into the modern era. In February 2022, it reopened as the "Empire of Cats" – an immersive habitat complex where Zoo visitors find themselves guests in the realm of animals. Now, a pride of lions roams a savannah-like landscape in front of the historic rocky backdrop, while a tiger watches every move made by human intruders in its territory. Before visitors can tear their eyes away from that impressive sight, they are already face to face

Crowds at the Panda Garden, 2017

Empire of Cats

Reintroduction of European bison in the Caucasus with the WWF

with a jaguar about to take a cool dip in its pool. The Empire of Cats also houses an in-depth exhibition where visitors of all ages can learn the secrets of these successful hunters and discover that even fearsome predators face dangers.

The Rhino Pagoda, which opened in 2023, pays a sophisticated modern homage to the historic buildings of Zoo Berlin. It is reminiscent of the once magnificent "Indian-style" Elephant Pagoda, where rhinoceroses were housed alongside elephants from the 1870s until the building was severely damaged in World War II and the ruins were removed in the 1950s. The new Rhino Pagoda houses Indian rhinoceroses, tapirs and warty pigs in an expansive swampy habitat. The impressive 25-metre tower draws attention even from a distance, and inside this monument to species conservation visitors can learn all about the diversity of nature and why we need to protect it. The exhibition provides answers to questions such as: What are the consequences of biodiversity loss? Are there still glimmers of hope? And: What can I do to help?

The Aquarium has also been modernised and equipped with many new didactic elements.

Despite the adverse circumstances brought about by the recent pandemic and war, the Tierpark has gone from strength to strength, offering the public many new worlds of experience. In 2020, the historic Alfred Brehm building was reopened as the Rainforest House, whisking visitors away to the endangered island paradises of

Rhino Pagoda

Southeast Asia. The "mountainous" Himalayan area opened in 2022, followed by the African Savannah in 2023, and Otter Island in 2024.

Securing a future: Berlin World Wild

Today, Zoo Berlin plays an important role in international species and nature conservation by actively combating global species extinction. As well as scientifically breeding endangered species in order to release them into their natural habitat or preserve them for future generations, the Zoo runs its own species conservation programme. Launched in 2020, Berlin World Wild follows a holistic approach, mobilising financial and human resources to help implement in situ conservation projects. With this programme, Zoo Berlin supports important conservation measures for endangered species around the globe. For instance, it channels support into snow leopard conservation in Central Asia, where these majestic animals are under threat. A large part of the funding also goes towards elephant conservation in southern Africa in order to secure a future for elephants in that region. Berlin World Wild embodies the Zoo's vision of preserving and protecting individual wild animals at its facilities while also actively contributing to wildlife conservation efforts around the world. The Zoo acts as a mediator between people and nature, raising awareness about the challenges posed by species extinction and giving visitors the opportunity to actively contribute to the preservation of endangered species. Berlin World Wild highlights Zoo Berlin's role as a pioneer in international nature conservation and sets a clear example for ways in which our planet's remarkable biodiversity can be preserved.

Zoo Berlin shareholders, Berliners from across the city, and state politicians are big fans and supporters of Berlin's three zoological facilities, which house a diversity of species that has developed over the course of many decades and is unique in Germany. In cooperation with the Leibniz Institute for Zoo and Wildlife Research, Berlin's natural history museum, and the zoological and veterinary facilities of two large universities, Zoo, Aquarium and Tierpark Berlin offer an unparalleled environment for education about and research into the diversity of animal life on our planet.

Visitor numbers confirm that not only can Berlin sustain two zoos – it *needs* two zoos. Even in an age when numerous nature documentaries are available at the touch of a button, zoos offer an unmatched experience of direct, authentic contact with animals of all shapes and sizes. Every year, roughly 3.5 million Zoo and Aquarium visitors and 1.5 million Tierpark visitors prove just how important such experiences still are today.

In 2024, the Zoological Garden of Berlin looks back on its 180 years of existence with pride, while remaining ready and willing to confront the darker chapters of its history with complete transparency. After all, only those who acknowledge all the facets of their own past can shape the future in a meaningful way.

Appendix

Notes

Rulers and their animals: The predecessors to Zoo Berlin (pp. 15–21)

1 References: Helen Cowie, *Exhibiting Animals in Nineteenth-Century Britain: Empathy, Education, Entertainment* (Basingstoke, 2014); Vernon N. Kisling, "Ancient Collections and Menageries", in id. (ed.), *Zoo and Aquarium History: Ancient Animal Collections to Zoological Gardens* (Boca Raton, 2001), 1–47; Eric Baratay and Elisabeth Hardouin-Fugier, *Zoo: Von der Menagerie zum Tierpark* (Berlin, 2000); Robert J. Hoage, Anne Roskell and Jane Mansour, "Menageries and Zoos to 1900", in Robert J. Hoage and William Deiss (eds.), *New Worlds, New Animals: From Menagerie to Zoological Park in the Nineteenth Century* (Baltimore, 1996), 8–18; Annelore Rieke-Müller and Lothar Dittrich, *Der Löwe brüllt nebenan: Die Gründung Zoologischer Gärten im deutschsprachigen Raum 1833–1869* (Cologne, 1998).

2 Harro Strehlow, "'Das merkwürdigste auf der Insel waren die lebendigen Thiere', Ein Besuch Christian Ludwig Brehms in der Menagerie auf der Pfaueninsel im Oktober 1832", *Bongo*, 24 (1994), 31–50; Caesar von der Ahé, "Die Menagerie auf der 'Königlichen Pfaueninsel': Der Ursprung des Berliner Zoologischen Gartens", *Mitteilungen des Vereins für die Geschichte Berlins*, 47/1 (1930), 1–24, here 5.

3 "Gehorsamster Reisebericht des Geheimen Medicinal-Raths Dr Lichtenstein", addressed to Education Minister Freiherr von Altenstein, GStA PK I. HA Rep. 76, Va, Section 2, Tit. X, No. 49, sheets 17–27.

4 Harri Günther and Sibylle Harksen, *Peter Joseph Lenné: Pläne* für Berlin (vol. ii of *Bestandskatalog der Lennépläne in der Plankammer der Staatlichen Schlösser und Gärten Potsdam-Sanssouci*; Berlin, 1989), 14ff.

5 Correspondence between Lichtenstein and the committee in GStA PK I. Rep. 76 Vc, Section 1, Tit 12, No 31823. Additional important information in: Rieke-Müller and Dittrich, *Gründung Zoologischer Gärten*, 53–63; Gary Bruce, *Through the Lion Gate: A History of the Berlin Zoo* (Oxford, 2017), 19–28.

A zoological garden by Berlin, 1844–1870 (pp. 23–39)

1 Inspection report, GStA PK I. HA Rep. 89, No. 31823, sheets 8–12.

2 Transcript in AZGB O 0/1/135.

3 Sovereign cabinet order of His Majesty the King, 16 Aug. 1843, AZGB O 0/1/135.

4 Inventory of animals at the Zoological Garden, mid-June 1845, AZGB S 9/4.

5 Martin Hinrich Lichtenstein, "Zoologischer Garten", *Vossische Zeitung*, 1 Aug. 1844.

6 "Der Zoologische Garten und das Publikum", *Vossische Zeitung*, 14 Aug. 1844.

7 Carl Rißle, "Der zoologische Garten der deutschen Hauptstadt", *Illustrierter Volkskalender der Berliner Bürger-Zeitung* (Berlin, 1875).

8 Drafts for a newspaper advertisement, June 1845 and 9 Aug. 1845, AZGB O 0/1/76.

9 "Zoologischer Garten", *Vossische Zeitung*, 29 July 1844. In a meeting on 14 June 1845, the board decided to compile a register of the existing animals, to be sold at the entrance (AZGB O 0/1/76).
10 M. Lichtenstein to the "Actien-Verein des zoologischen Gartens bei Berlin", 5 July 1845, and minutes of a board meeting on 12 July 1845, AZGB O 0/1/76.
11 "Bescheidene Anfrage", *Vossische Zeitung*, 12 July 1845; M. Lichtenstein, "Erwiederung", *Vossische Zeitung*, 12 July 1845.
12 Board meeting minutes, 19 July 1845, AZGB O 0/1/76.
13 Report by the Royal Prussian Consulate in Algiers on the delivery of wild animals by General Yusuf to Consul Gustav Lichtenstein in Marseilles, 10 Aug. 1847, AZGB N 10/1.
14 Board meeting minutes, 22 Sept. 1845, AZGB O 0/1/76, and animal inventory of Aug. 1846 plus a supplement the following month, LAB E Rep. 200-20, No. 425.
15 Board meeting minutes, 28 Aug. 1845, AZGB O 0/1/76.
16 Preliminary financial statement for the year 1846 of the "Actien-Verein des zoologischen Gartens bei Berlin", AZGB O 0/1/137.
17 M. Lichtenstein to the "Actien-Verein des zoologischen Gartens bei Berlin", 5 July 1845, AZGB O 0/1/76.
18 Statutes of the "Actien-Vereins des zoologischen Gartens bei Berlin" and the associated zoological society, 28 Feb. 1845, AZGB O 0/1/31.
19 F. Eichhorn to Friedrich Wilhelm IV, 27 Mar. 1845, GStA PK I. HA Rep. 89, No. 31823, sheets 44–51; sovereign cabinet order of His Majesty the King, 13 Feb. 1843, AZGB O 0/1/135.
20 Board meeting minutes, 21 June 1845, AZGB O 0/1/76.
21 Ludwig Heck, "Fünfundsiebzig Jahre Zoologischer Garten", *Leipziger Illustrirte Zeitung*, 31 July 1919.
22 List of shareholders of 28 May 1859, AZGB O 0/1/134.
23 Board of the Actien-Verein to finance minister Flottwell, 30 July 1845, AZGB O 0/1/76.
24 Feuilleton of *Berliner Figaro*, 28 Aug. 1845, 799ff.
25 E. v. Puttkamer to H. Lichtenstein, 10 May 1843, AZGB O 0/1/73.
26 Inventory of animals at the Zoological Garden, AZGB S 9/4.
27 Ludwig Heck, "Fünfundsiebzig Jahre".
28 Rißle, "Der zoologische Garten", 108–12.
29 Heinz-Georg Klös, "Tierhaltung im alten Berlin", part two, *Bongo* 14 (1988), 105–14, here 106ff.; board meeting minutes, 18 Feb 1863, O 0/1/146.
30 Minutes of the general meeting, 1 June 1852, AZGB O 0/1/62.
31 Board meeting minutes, summer 1857, AZGB O 0/1/146.
32 Zoologischer Garten Berlin AG, statute of the "Actien-Verein des Zoologischen Gartens zu Berlin", 14 May 1869.
33 Ludwig Heck, *Führer durch den Berliner Zoologischen Garten* (Berlin, 1901), 5; Adolf Heilborn, *Zoo Berlin 1841–1929: Zur Geschichte des Zoologischen Gartens zu Berlin*, ed. the board of the "Actien-Verein des Zoologischen Gartens zu Berlin" (Berlin, 1929), 24.
34 Heinz-Sigurd Raethel, "Die Großkatzenhaltung im Berliner Zoo in den Jahren 1844 bis 1945", *Bongo* 40 (2010), 38–59, here 41; Royal Police Office to the Zoo, 15 June 1858, AZGB O 0/1/136.
35 "Affenjagd an der Gedächtniskirche", *Spandauer Volksblatt*, 21 Aug. 1954.
36 "Gorilla verfolgte Familie im Berliner Zoo", *B.Z.*, 17 Sept. 1974.
37 Zoologischer Garten Berlin annual report for 1979 (Berlin, 1980).
38 Werner Synakiewicz, *Berliner Zoo: Die schönsten Geschichten* (Berlin, 2004), 35–7.
39 Copy in AZGB S 15/17.
40 Draft letter from the chairman of the board to the Charlottenburg municipal authorities, 4 May 1883, AZGB O 0/1/136.
41 Recollections of Willi Kettner, AZGB S 15/26.
42 Telegrams, correspondence and letters, AZGB S 15/17 and O 0/1/115.

The early years of the empire: A city and zoo in bloom, 1870–1900 (pp. 41–65).

1 Heinz-Georg Klös and Ursula Klös (eds.), *Der Berliner Zoo im Spiegel seiner Bauten 1841–1989: Eine baugeschichtliche und denkmalpflegerische Dokumentation über den Zoologischen Garten Berlin* (Berlin, 1990), 49.
2 Carl Rißle, "Der zoologische Garten der deutschen Hauptstadt", *Illustrierter Volkskalender der Berliner Bürger-Zeitung* (Berlin, 1875).
3 Judicial counsellor Stämmler to the Zoo, 4 May 1880, AZGB O 0/1/147.
4 Dossier in GStA PK I HA, Rep. 89, No. 31823.
5 Klös and Klös, *Der Berliner Zoo im Spiegel seiner Bauten*, 58.
6 Albert Hofmann, "Die Umwandlung und die Neubauten des Zoologischen Gartens in Berlin", *Deutsche Bauzeitung* 36/22 (1902), 18.
7 F. Lichterfeld, "Die Dickhäuter des berliner Zoologischen Gartens und ihr Palast", *Leipziger Illustrirte Zeitung*, 15 Aug. 1874.
8 Hofmann, "Die Umwandlung und die Neubauten", 70ff.

9 Hans Frädrich, "Zoologische Gärten heute: Zwischen Museen, Naturschutz- und Freizeiteinrichtungen am Beispiel des Berliner Zoos", *Museumsjournal* 8/3 (1994), 19–22, here 19.
10 Ludwig Heck, "Der Brand des Raubthierhauses im berliner Zoologischen Garten", *Berliner Illustrirte Zeitung*, 26 Oct. 1899.
11 "Fr.", "Der Zoologische Garten in Berlin", *Gartenlaube* 16 (1873).
12 F. Lichterfeld, "Der Zoologische Garten in Berlin", *Leipziger Illustrirte Zeitung*, 4 May 1872.
13 "Das Elefantenskelet im berliner Zoologischen Garten", *Leipziger Illustrirte Zeitung*, 17 Mar. 1883.
14 Directorate of the Pferde-Eisenbahn-Gesellschaft to the Zoo, 24 Feb. 1875, and further correspondence relating to the creation of a depot, AZGB O 0/1/83.
15 Construction drawings and layout maps in LAB A Rep. 010-02, Nos. 31588 and 31590. The Pferde-Eisenbahn-Gesellschaft published the tram timetable in the *Vossische Zeitung* of 3 July 1875. I am indebted for this information to the archives of the Berliner Verkehrsbetriebe (BVG).
16 Helmut Bombe, *Geschichte der ersten Pferdeeisenbahn Deutschlands* (no place or date). The manuscript was kindly provided by the archives of the Berliner Verkehrsbetriebe (BVG).
17 Directorate of the Berliner Stadt-Eisenbahn to the Zoo, 2 Nov. 1881, AZGB O 0/1/69.
18 *Zeitschrift für Bauwesen* 35/10–12 (1885), 504.
19 Supervisory board meeting minutes, 13 Dec. 1910, AZGB O 0/2/23.
20 Correspondence between the ministry and the Pferde-Eisenbahn-Gesellschaft, AZGB O 0/1/83.
21 Correspondence between the ministry and the Pferde-Eisenbahn-Gesellschaft, AZGB O 0/1/83.
22 Ministry of the Interior, *Statistisches Jahrbuch der Stadt Berlin* 16/17 (the years 1889 and 1890).
23 Rudolf Leuckert to the Zoo, 20 May 1888, AZGB O 0/2/13.
24 Ludwig Heck, *Heiter-ernste Lebensbeichte: Erinnerungen eines alten Tiergärtners* (Berlin, 1938), 195.
25 An extensive description of the buildings at that time can be found in Hofmann, "Die Umwandlung und die Neubauten".
26 *Die Thierwelt im Zoologischen Garten von Berlin, geschildert von Dr. Bodinus, Dr. Max Schmidt und Dr. L. Heck, Directoren des Gartens* (Berlin, ca. 1897): 54ff.
27 Hofmann, "Die Umwandlung und die Neubauten".
28 Michael H. Robinson, foreword in Robert J. Hoage, William A. Deiss and Herman Reichenbach (eds.), *New Worlds, New Animals: From Menagerie to Zoological Park in the Nineteenth Century* (Baltimore, 1996), vii–xi.
29 Information on the "ethnological exhibitions" is provided primarily by: Anne Dreesbach, *Gezähmte Wilde: Die Zurschaustellung "exotischer" Menschen in Deutschland 1870–1940* (Frankfurt/M., 2005); Pascal Blanchard et al. (eds.): *MenschenZoos: Schaufenster der Unmenschlichkeit* (Hamburg, 2012), with many individual contributions; Balthasar Staehelin, *Völkerschauen im Zoologischen Garten Basel 1879–1935* (Beiträge zur Afrikakunde, 11; Basel, 1939); Stefanie Wolter, *Die Vermarktung des Fremden: Exotismus und die Anfänge des Massenkonsums* (Frankfurt/M., 2005). For information about Carl Hagenbeck's company, I recommend: Hilke Thode-Arora, *Für fünfzig Pfennig um die Welt: Die Hagenbeckschen Völkerschauen* (Frankfurt/M., 1989); Lothar Dittrich and Annelore Rieke-Müller, *Carl Hagenbeck (1844–1913): Tierhandel und Schaustellungen im Deutschen Kaiserreich* (Frankfurt/M., 1998). For specific information about the exhibitions of Samoan groups: Hilke Thode-Arora (ed.), *From Samoa With Love? Samoa-Völkerschauen im deutschen Kaiserreich* (Munich, 2014). Ursula Klös created the first and, to date, most detailed account of the "ethnological exhibitions" at Zoo Berlin: "Völkerschauen im Zoo Berlin zwischen 1878 und 1952", *Bongo* 30 (2000), 33–82.
30 Board meeting minutes, 5 Dec. 1877, AZGB O 0/1/104.
31 Board meeting minutes, 2 Oct. 1878, AZGB O 0/1/104.
32 France Rivet, *In the Footsteps of Abraham Ulrikab: The Events of 1880–1881* (Gatineau, 2016).
33 For details about the origins of the group: Gabriele Eißenberger, *Entführt, verspottet und gestorben: Lateinamerikanische Völkerschauen in deutschen Zoos* (critical and self-critical research papers on the "Third World" published by the Arbeitsstelle Dritte Welt department at TU Berlin; Frankfurt/M., 1996), 145–75.
34 Dreesbach, *Gezähmte Wilde*, 223.
35 Klös, "Völkerschauen", 43.
36 C. Hagenbeck to the Zoo, 28 Sept. 1881, AZGB O 0/1/69.
37 Eißenberger, *Entführt, verspottet und gestorben*, 166–71.
38 Board meeting minutes, 8 Jan. 1883, AZGB O 0/1/141.
39 Zoologischer Garten Berlin annual report for 1926 (Berlin, 1927).
40 C. Marquardt to the Zoo, 7 Dec. 1899, AZGB O 0/1/42.
41 Contract between the Zoo and C. Marquardt, AZGB O 0/1/42.
42 Press report "An die verehrte Redaktion", AZGB O 0/1/42.
43 Report on a business trip to Rome relating to the

Tripoli show, probably written by incumbent administrative director Clemens Freyer, 6 May 1927, AZGB O 0/1/210.

44 Account of earnings and losses for individual exhibitions from 1926 to 1928, AZGB O 0/1/114.

45 Annual report of the "Actien-Verein des Zoologischen Gartens Berlin" for 1927 (Berlin, 1928).

46 Frankfurt Zoo to Zoo Berlin, 29 Dec. 1926 and a carbon copy of the reply from 30 Dec. 1926, AZGB O 0/1/130.

47 Correspondence and draft contract from F. W. Siebold, 29 Jan 1931, AZGB S 15/17.

Global fame and global crises, 1900–1932 (pp. 67–91)

1 *Berliner Illustrirte Zeitung*, 15 May, 21 July, 18 Sept. and 18 Dec. 1902.

2 "Dies und das", *Die Woche*, 3 Feb. 1900; *Berliner Illustrirte Zeitung*, 1 Feb. 1900.

3 "Der Kaiser bei Hagenbeck", *Berliner Tageblatt*, evening edition, 18 June 1912.

4 "Actien-Verein des Zoologischen Gartens" to Friedrich von Hollmann, 10 Apr. 1911, GStA PK I HA, Rep. 89, royal privy cabinet, latter period, No. 31824, sheets 48–52.

5 Lothar Dittrich and Annelore Rieke-Müller, *Carl Hagenbeck (1844–1913): Tierhandel und Schaustellungen im Deutschen Kaiserreich* (Frankfurt/M., 1998), 272ff.

6 Heinz-Georg Klös and Ursula Klös (eds.), *Der Berliner Zoo im Spiegel seiner Bauten 1841–1989: Eine baugeschichtliche und denkmalpflegerische Dokumentation über den Zoologischen Garten Berlin* (Berlin, 1990), 167ff.; correspondence with the Prussian finance ministry, the Berlin authorities and the shareholder groups in AZGB O 0/1/132.

7 "Restaurations- und Saalneubauten im Berliner Zoologischen Garten", *Berliner Architekturwelt* 7/14 (1911), 253–91.

8 Bernd Weiler, *Die Ordnung des Fortschritts: Zum Aufstieg und Fall der Fortschrittsidee in der "jungen" Anthropologie* (Bielefeld, 2006), 541; Harro Strehlow, "Andenken an das Berliner Aquarium", *Bongo* 41 (2011), 75–86.

9 Supervisory board meeting minutes, 30 Jan. 1909, AZGB O 0/2/23; Zoo to state commissioner, 30 Sept. 1907, AZGB O 0/1/144.

10 "Der Oberbürgermeister als Elbfischer", *Berliner Volks-Zeitung*, morning edition, supplement, 19 Aug. 1933; correspondence with the court and the approval of the king in 1910 in GStA PK I HA, Rep. 89, royal privy cabinet, latter period, No. 31824, folios 38–41; Oskar Heinroth, "25 Jahre Berliner Aquarium im Zoologischen Garten" (Berlin, 1938; AZGB S 2/6); Zoologischer Garten Berlin annual report for 1910 (Berlin, 1911).

11 Supervisory board meeting minutes, 13 Dec. 1910, AZGB O 0/2/23.

12 Klös and Klös, *Der Berliner Zoo im Spiegel seiner Bauten*, 163–6.

13 Katharina Heinroth, *Oskar Heinroth* (Große Naturforscher, 34; Stuttgart, 1971), 103.

14 Heinz-Georg Klös and Ursula Klös, "Dr. Oskar Heinroth zum 125. Geburtstag", *Bongo* 27 (1997), 3–22.

15 Karl Schulze-Hagen and Timothy R. Birkhead, "The Ethology and Life History of Birds: The Forgotten Contributions of Oskar, Magdalena and Katharina Heinroth", *Journal of Ornithology* 156 (2015), 9–18.

16 Supervisory board meeting minutes, 18 Aug. 1914, AZGB O 0/2/22.

17 Zoologischer Garten Berlin annual reports for 1914 to 1918 (Berlin, 1916, 1917, 1918 and 1919).

18 Zoologischer Garten Berlin annual report for 1918 (Berlin, 1919).

19 Minutes of a meeting of the finance commission, 14 Mar. 1919, AZGB O 0/2/22.

20 K. Heinroth, *Oskar Heinroth*, 105.

21 Curriculum vitae and questionnaires of Lutz Heck, Federal Archives, R 9361-V/ 5953, digital files Nos. 144 and 148.

22 Zoologischer Garten Berlin annual report for 1919 (Berlin, 1920).

23 Supervisory board meeting minutes, 30 Aug. 1920, AZGB O 0/2/22.

24 Supervisory board meeting minutes, 23 Dec. 1920, AZGB O 0/2/22.

25 "Der Zoo in Gefahr", *Der Tag*, supplement, 27 July 1922.

26 Correspondence on fundraising and police authorisations in GStA PK I. HA Rep. 191, No. 4296.

27 Zoologischer Garten Berlin annual report for 1923 (Berlin, 1924).

28 Letter from the Prussian prime minister to the Prussian finance ministry, 23 Aug. 1924, and the ministerial resolution of the finance ministry, 26 Jan. 1934; Zoo to the Prussian finance ministry, 1 Aug. 1940, GStA PK I. HA Rep. 151, finance ministry, No. 2496, sheets 190–201.

29 "Ach, das Beest mag lieber wo anders verrecken!", 4th supplement of the 8 o'clock evening edition of the *National-Zeitung*, 3 Nov. 1928; Zoologischer Garten Berlin annual report for 1895 (Berlin, 1896).

30 Report regarding the land at the edge of Zoo Berlin, AZGB O/0/3/10.

31 Supervisory board meeting minutes, 30 Aug. 1920, AZGB O 0/2/22; Zoologischer Garten Berlin annual report for 1920 (Berlin, 1921).
32 Harro Strehlow, "Die Teneriffa-Schimpansen und der Zoologische Garten Berlin", *Bongo* 25 (1995), 47–52.
33 Invitation to a press preview, 2 July 1924, AZGB N 2/4; "Abessinische Geschenke", *Die Woche*, issue 46, (1923), 1047ff.
34 "Z.B. Exotischer Besuch", *Berliner Morgen-Zeitung*, 21 May 1928.
35 Bernhard Blaszkiewitz, *Elefanten in Berlin* (Berlin, 2008), 13.
36 "K.", "Das neue Affenpalais Berlins", *Der Tag*, night-time edition, 18 Sept. 1954.
37 Zoologischer Garten Berlin annual report for 1926 (Berlin, 1927); extract from a speech by privy councillor Prof. Heck in June 1926, AZGB S 2/13.
38 Printed guide to the Zoo Berlin in 1941 (Berlin, 1941), 14.
39 Correspondence with the municipal authorities, LAB A Rep. 001-002, No. 899.
40 Appeal and correspondence, AZGB O 0/1/81 and LAB A Rep. 001-002, No. 899.
41 Franz Schirner, "Sind zwei Berliner Zoologische Gärten das Richtige?", *Neue Zeit*, 11 Sept. 1929.
42 "Hagenbeck in Berlin", *Deutsche Tageszeitung*, 27 Oct. 1929.
43 "Ein Hagenbeck-Tierpark für Berlin", *Berliner Tageblatt*, 24 Oct. 1929; "Umwälzende Berliner Projekte", *Neuköllner Tageblatt*, 26 Oct. 1929.
44 "Zoo-Verlegung nach Neu-Cladow?", *Vossische Zeitung*, 26 June 1929.
45 "Vollständige Umgestaltung des Westens", *Vossische Zeitung*, 25 Nov. 1928; "Wohin mit dem Zoo?", 1st supplement to *Vorwärts*, 13 June 1929.
46 According to Spandau municipal planning and building officer Ludorf in "Um den Zoologischen Garten", *Spandauer Zeitung*, 7 June 1929.
47 Schmargendorf house and property owners' association to the directorate of the Zoologischer Garten, 10 July 1929, AZGB O 0/1/81.
48 "Volkstierpark Wuhlheide?", *Tageblatt*, 22 Oct. 1929.
49 Memo by H. Ammon, 25 Feb. 1930, AZGB O 0/1/81.
50 Speech by Ludwig Heck, AZGB O 0/1/81.
51 "Zoo und Volkstierpark", *Der Berliner Westen*, 7 Dec. 1929.
52 Supervisory board meeting minutes, 22 Feb. 1933, AZGB O 0/2/2, sheet 30.
53 Correspondence with the Ruhe company in Hanover, 1931, AZGB O 3/2/1.

In thrall to the Swastika:
Zoo Berlin during the Third Reich, 1933–1945
(pp. 93–131)

1 Supervisory board meeting minutes, 22 May 1933, AZGB O 0/2/2.
2 Zoologischer Garten Berlin annual report for 1933 (Berlin, 1934).
3 Ludwig Heck, *Heiter-ernste Lebensbeichte: Erinnerungen eines alten Tiergärtners* (Berlin, 1938), 373.
4 Lutz Heck, *Waidwerk mit bunter Strecke* (Berlin, 1968), 67.
5 For example, Lutz Heck, *Der deutsche Edelhirsch: Ein Lebensbild* (Berlin, 1935).
6 Yearbook of the Fachschaft Deutsche Bracken, 1935/36.
7 Zoologischer Garten Berlin, old animal register and lists of Hermann Göring's lions plus their offspring, AZGB N 5/13.
8 Manuscript by an unknown author about the reintroduction of European bison in the Schorfheide, AZGB S 15/3.
9 Lutz Heck, *Auf Urwild in Kanada* (Berlin, 1937).
10 Lutz Heck, *Auf Tiersuche in weiter Welt* (Berlin, 1941), 189ff.
11 Carbon copy of a letter from K. Heinroth to lawyer E. Leser on 28 Oct. 1948, AZGB N 4/4.
12 Zoologischer Garten Berlin annual report for 1933 (Berlin, 1934).
13 Monika Schmidt has rendered an outstanding service with her reconstruction of the events relating to the Jewish supervisory board members and several Jewish shareholders and their subsequent fates, for example in her book *Die jüdischen Aktionäre des Zoologischen Gartens zu Berlin: Namen und Schicksale* (Berlin, 2014) and in her article "Die 'Arisierung' des Berliner Zoologischen Gartens", *Jahrbuch für Antisemitismusforschung* 12 (Berlin, 2003), 210–29.
14 Supervisory board meeting minutes, 21 Feb. 1933, AZGB O 0/2/2, sheet 29.
15 G. Siegmann to K. Gelpcke, 1 Mar. 1933, AZGB O 0/2/20.
16 Supervisory board meeting minutes, 6 Apr. 1933, AZGB O 0/2/2, sheets 35–6.
17 Memo for district president Zachariae, GStA PK I. HA Rep. 151, No. 2496, sheet 23.
18 Carbon copy of a letter from Karl Gelpcke to Adolf von Mecklenburg, 8 Apr. 1933, AZGB O 0/2/20.
19 "Zoo-Generalversammlung", *Berliner Börsenzeitung*, 29 Apr. 1933.
20 All quotes from the supervisory board meeting minutes of 3 May 1933, AZGB O 0/2/2.
21 Memo, 11 May 1933, AZGB O 0/2/2.

22 Supervisory board meeting minutes, 22 May 1933, AZGB O 0/2/2.
23 Supervisory board meeting minutes, 13 June 1933, AZGB O 0/2/2.
24 Schmidt, *Die jüdischen Aktionäre*, 26.
25 Memo, 9 June 1933, GStA PK I. HA Rep. 151, finance ministry, No. 2496, sheets 27–9.
26 Supervisory board meeting minutes, 6 Sept. 1933, AZGB O 0/2/2.
27 Angelika Königseder, *Recht und nationalsozialistische Herrschaft: Berliner Anwälte 1933–1945; ein Forschungsprojekt des Berliner Anwaltsvereins e.V.* (Bonn, 2001), 214.
28 Supervisory board meeting minutes, 2 Nov. 1933, AZGB O 0/2/2.
29 Minutes of the general meeting, 1934, AZGB O 0/3/2.
30 Zoologischer Garten Berlin annual report for 1935 (Berlin, 1936).
31 Hermann Steinmetz, "Tiere im Olympischen Dorf", *Geflügel-Börse*, 21 July 1936.
32 Zoologischer Garten Berlin annual report for 1936 (Berlin, 1937); "DrWC", "Zoowärter lernen Englisch", *Berliner Illustrierte*, night-time edition, 22 Apr. 1936.
33 "CH", "Der Wunschtraum: ein deutscher Zoo; Ein Gespräch mit Dr. Lutz Heck", *Völkischer Beobachter*, Berlin edition, 8 June 1934.
34 "Berlins Zoo wird erweitert", *Berliner Nachtausgabe*, 15 October 1935.
35 Correspondence between all parties concerned in GStA PK I. HA, Rep. 151, 2500; transcript of the supervisory board meeting of 24 Aug. 1935 in GStA PK I. HA, Rep. 151, 2496, sheets 93–4.
36 "Berlin bekommt einen Deutschen Zoo", *Berliner Lokal-Anzeiger*, 23 Dec. 1936.
37 Kai Artinger, "Lutz Heck: Der 'Vater der Rominter Ure': Einige Bemerkungen zum wissenschaftlichen Leiter des Berliner Zoos im Nationalsozialismus", *Der Bär von Berlin: Jahrbuch des Vereins für die Geschichte Berlins* 23 (1994), 127. Perhaps, however, this is also a reference to the flags mentioned in the *Völkischer Beobachter*. The Reich flag introduced by the Nazis had a swastika at the centre.
38 Press tour of the domestic animal "homestead" on Whit Sunday in 1937, AZGB O 0/1/15.
39 Membership card in the Berlin Document Center of the Federal Archives.
40 For example, L. Heck to H.-G. Klös, 29 Jan. 1969, AZGB N 5/18.
41 Form for admission to the Reich Chamber of Literature, undated, Federal Archives, NS 21/1543.
42 Copy of a letter from L. Heck to Alfred Zoll (Hellabrunn zoo, Munich), 24 Nov. 1939, AZGB N 2/2; Lutz Heck, "Über die Neuzüchtung des Ur oder Auerochs", *Berichte der Internationalen Gesellschaft zur Erhaltung des Wisents* 3/4 (1936), 224–94, here 235.
43 Heinz Heck, "Über die Rückzüchtung des Urs", *Das Tier und wir* (1936), Oct., 8–14, here 8.
44 Andreas Gautschi, *Der Reichsjägermeister: Fakten und Legenden um Hermann Göring* (Melsungen, 2010), 79.
45 Lutz Heck, "Über die Neuzüchtung des Urs oder Auerochs", in the official catalogue of the International Hunting Exhibition in Berlin, 2 to 28 Nov. 1937 (Berlin, 1937), 428–30.
46 Lutz Heck, *Auf Tiersuche*, 195.
47 Lutz Heck, "Über die Neuzüchtung des Urs oder Auerochs", 294.
48 Lutz Heck, *Waidwerk*, 79; Hermann Steinmetz, "Berlin", *Der Zoologische Garten* 11/1–2 (1939), 32–8, here 35.
49 Lutz Heck, "Die Neuzüchtung des Auerochsen", *Wild und Hund* 37 (1939), 535–7, here 537.
50 Zoologischer Garten Berlin, old animal register, information on breeding attempts.
51 Heinz Heck, "Der Ur – ein vergessenes Tier", *Das Tier und wir* (1934), issue 3, 1–16, here 16; also Heinz Heck, "Über die Rückzüchtung des Urs", 13.
52 Lutz Heck, "Letzte Urwaldtiere aus deutscher Vorzeit", *Atlantis: Länder, Völker, Reisen* 4/10 (1932), 577–83; also "Die Neuzüchtung des Auerochsen", 535–37 and *Auf Tiersuche*, 195. At the International Hunting Exhibition, a frieze featuring a verse from the epic poem adorned the aurochs stand.
53 Information about when Lutz Heck was appointed head of the nature conservation department is inconsistent. Heck himself later testified that he had remained loyal to the Zoo. However, various pieces of evidence suggest that he was responsible for nature conservation by autumn 1938 at the latest and that he led the new nature conservation department from spring 1941. See the copy of a letter from department head Ebert at the Reich Forestry Office to the head of the Reichsbund für Biologie Dr W. Greite on 5 Feb. 1941, Federal Archives, NS 21/1543; supervisory board meeting minutes, 19 July 1938, AZGB O 0/2/2. See also Lutz Heck, *Waidwerk*, 71; "Berlin", *Der Zoologische Garten* 12/1 (1940), 88; Frank Uekötter, *The Green and the Brown: a History of Conservation in Nazi Germany* (Studies in Environment and History; Cambridge, 2006), 72.
54 Lutz Heck, *Waidwerk*, 71.
55 Heinrich Dathe and Tierpark Berlin (eds.), *Internationales Zuchtbuch für Auerochsen: International Studbook of the Aurox-like Domestic Cattle*, 1 (Berlin, 1980), 5ff.

56 Letter from H. Ammon to "Zoo Elberfeld", 22 Mar. 1938 in Ulrich Schürer and Gerd Schmerenbeck, *100 Jahre Zoo Wuppertal: Chronik*, ed. Informationsamt der Stadt Wuppertal and Zoologischer Garten Wuppertal (Wuppertal, 1981); and Schmidt, *Die jüdischen Aktionäre*, 33; Wilhelm Seiffge to Zoo Berlin, 23 June 1939, AZGB O 0/1/56.
57 Supervisory board meeting minutes, 29 Mar. 1938 and 17 May 1938, AZGB O 0/2/2.
58 Supervisory board meeting minutes, 8 Nov. 1938, AZGB O 0/2/2.
59 Supervisory board meeting minutes, 16 Dec. 1938, AZGB O 0/2/2.
60 *Picture Post* 152, after June 1939.
61 Minutes of the general meeting in 1939, AZGB O 0/3/4.
62 For the history of some Jewish shareholders and their fate during the Nazi regime see Schmidt, *Die jüdischen Aktionäre*.
63 Supervisory board meeting minutes, 19 July 1938, AZGB O 0/2/2.
64 Supervisory board meeting minutes, 16 Dec. 1938, AZGB O 0/2/2. The report on the audit of the annual accounts on 31 Dec. 1941 (GStA PK, I. HA, Rep. 151, No. 2497) refers to 80 shares. It is uncertain whether these are the same shares. In the following year the Zoo recorded 6,550.20 Reichsmarks from their sale, which would suggest a relatively small number of shares, see Report on the audit of the annual accounts on 31 Dec. 1942, 1943, AZGB O S 17/2.
65 Speech by H. Ammon at the AGM in May 1940, AZGB O 0/3/13.
66 Speech by H. Ammon at the general meeting in 1942, AZGB O 0/3/12.
67 Correspondence with the Prussian state bank in 1944, AZGB O 0/1/43.
68 Supervisory board meeting minutes, 16 Dec. 1939, AZGB O 0/2/2; copy of letters from L. Heck to various zoos, 14 Sept. 1939, AZGB S 15/30.
69 Report on the audit of the annual accounts on 31 Dec. 1939, AZGB S 17/1.
70 Speech by H. Ammon at the AGM in 1940, AZGB O 0/3/13.
71 Hermann Steinmetz, "Berlin", *Der Zoologische Garten* 13/3–4 (1941), 307–13.
72 Supervisory board meeting minutes, 30 July 1940, AZGB O 0/2/2.
73 Ibid.; Steinmetz, "Berlin", *Der Zoologische Garten* 14/5–6 (1942), 277–85.
74 Drawings and applications in LAB A Rep. 032-08, No. 293.
75 Manuscript by Lutz Heck, "Der Luftangriff vom 22. November 1943 auf den Zoologischen Garten", AZGB N 4/5; Lutz Heck, *Tiere – mein Abenteuer: Erlebnisse in Wildnis und Zoo* (Vienna, 1954), 102.
76 Drawings and applications in LAB A Rep. 032-08, No. 297; "K. Kr.", "Im Gämsenberg ist ein Luftschutzraum", *Berliner Börsen-Zeitung*, 22 Nov. 1939.
77 Cf. Mayumi Itoh, *Japanese Wartime Zoo Policy: The Silent Victims of World War II* (New York, 2010).
78 Lutz Heck, *Tiere – mein Abenteuer*, 97–9; Steinmetz, "Berlin", *Der Zoologische Garten* 14/5–6 (1942), 277–85.
79 Speech by L. Heck at the AGM in 1940, AZGB O 0/3/13; Lutz Heck, *Tiere – mein Abenteuer*, 99; Steinmetz, "Berlin" (1942); note on the annual report for 1941, AZGB O 0/3/12.
80 Recollections of Elisabeth Johst, AZGB S 15/27.
81 Speech by L. Heck at the AGM in 1940, AZGB O 0/3/13, and supervisory board meeting minutes, 28 Mar. 1940, AZGB O 0/2/2. The orders from Jan. 1940 can be found in Federal Archives R2/4749.
82 Draft letter from J. Popitz to the Reich Ministry of Armaments and War Production, 12 Oct. 1940; letter from the General Plenipotentiary for Regulation of the Construction Industry, Inspector General Todt, to J. Popitz, 16 Oct. 1940, GStA PK I. HA Rep. 151, Finance Ministry, No. 2496, sheets 219–25.
83 "'Der Urwald ruft' Kolonialkunst-Ausstellung im Zoologischen Garten", *Berliner Lokalanzeiger*, 6 Apr. 1933.
84 Press release, 29 June 1934, AZGB O 0/1/15; "Sensation im Affenpalmenhaus", *Völkischer Beobachter*, 13 June 1937.
85 Correspondence and invitations to meetings from 1939 to 1941 in AZGB S 15/29.
86 Files on "colonial hunting and nature conservation" in AZGB S 15/14, S 15/15 and S 15/16.
87 This "Dr Müller" was probably the subsequent zoo director in occupied Poznań. A report on the state of the Warsaw Zoological Garden by Dr Müller around Sept. 1939, AZGB S 15/30.
88 Gary Bruce, *Through the Lion Gate: A History of the Berlin Zoo* (Oxford, 2017), 164, with reference to the memoirs of the zoo director's wife Antonina Żabińska.
89 Supervisory board meeting minutes, 16 Dec. 1939, AZGB O 0/2/2; draft manuscript of a speech by L. Heck for the 1940 AGM, AZGB O 0/3/13; Zoologischer Garten Berlin, old animal register.
90 Honoured as "Righteous Among the Nations" on 21 Sept. 1965 by the Yad Vashem World Holocaust Remembrance Center, https://www.yadvashem.org/righteous/stories/zabinski.html (accessed 2 May 2024).

91 Supervisory board meeting minutes, 30 July 1940, AZGB O 0/2/2.

92 Andreas Gautschi, *Walter Frevert: Eines Waidmanns Wechsel und Wege* (Melsungen, 2005); Andreas Gautschi, *Der Reichsjägermeister*; Heinrich Rubner, *Deutsche Forstgeschichte, 1933–1945: Forstwirtschaft, Jagd und Umwelt im NS-Staat* (St. Katharinen, 1997).

93 Ulrich Scherping, "Bialowies wieder in deutscher Verwaltung", *Wild und Hund* (1941), 37/38, 317–25.

94 Copy of an agreement between the office of the Reich Commissioner for the Consolidation of German Nationhood and the Reich Forestry Minister as the supreme nature conservation authority on the execution of a discussion of 20 Mar. 1942, 11 May 1942, Federal Archives, R 49/2066.

95 Lutz Heck, "Behördliche Landschaftsgestaltung im Osten", *Neues Bauerntum mit Landbaumeister* 6/34 (1942), 213–15.

96 Speeches by H. Ammon and L. Heck at the 1942 AGM, AZGB O 0/3/12; Lutz Heck, *Waidwerk*, 83. There is clear evidence that Lutz Heck was in Białowieża in Jan. 1942. I am indebted for this information to Helmut Suter and Andreas Gautschi. It seems Heck was also there in winter 1943/44 (Gautschi, *Frevert*, 83).

97 Correspondence with the British commander in the Tiergarten district Lt. Col. Nunn in Dec. 1945, AZGB S 15/17; old animal register, "Panjepferde" index card.

98 Lutz Heck, "Askania Nova – heute", *Wild und Hund* 2 (1943), 12–15; Lutz Artinger, *Heck*, 132 and footnote 62; "Extract from Soviet War News: Reservation Park at Askanya Nova Devastated – Culprits Named", *Journal of the Society for the Preservation of the Fauna of the Empire* (London, 1944), 37ff.

99 Copies of correspondence between Lutz Heck and Otto Antonius (Schönbrunn Zoo), 1943, AZGB N 5/23.

100 Speeches by H. Ammon at the 1940 and 1941 AGMs, AZGB O 0/3/11 and O 0/3/12; copy of a letter from Lutz Heck to military administration Wehrbezirks-Kommando VIII, 4 Dec. 1943, AZGB O 0/1/66.

101 Reich Forestry Minister, RV 0080/1942.31, regarding securing workers for the Zoo, 3 June 1941, AZGB S 15/30.

102 Supervisory board meeting minutes, 16 Dec. 1939, AZGB O 0/2/2.

103 Minutes of the meeting of 13 February 1941, AZGB S 15/30.

104 Circular letter written by L. Heck to the zoological gardens, 22 Feb. 1945; letter from fish wholesaler Fisch-Grosshandel H. D. Petersen to L. Heck, 8 March 1945, AZGB O 0/1/88.

105 Zoologischer Garten Berlin annual report for 1940; a speech by Hans Ammon at the 1941 AGM, AZGB O 0/3/11; report on the audit of the annual accounts on 31 Dec. 1942, 1943, AZGB S 17/1.

106 Reich criminal investigation office to the Zoo, 8 Aug. 1944, AZGB O 0/1/43.

107 Speech by H. Ammon at the 1941 AGM, AZGB O 0/3/11.

108 Interview with Elisabeth Johst, University of Ohio Library, Cornelius Ryan Collection of World War II Papers, Box 59, Folder 34; the recollections of Elisabeth Johst, AZGB S 15/27.

109 Suggested by the bomb damage register, "Raubtierhaus" index card, AZGB O 0/1/54.

110 Planning permission and correspondence in Federal Archives, R 4606/4886.

111 Speech by H. Ammon at the 1942 AGM, AZGB O 0/3/12. In his speech, Lutz Heck also referred to "the white Russians" and "the French".

112 Memo of 8 Jan. 1944, AZGB O 0/1/41.

113 Harald Reissig, "Zoologischer Garten Hardenbergplatz 1", in Helmut Engel, Stefi Jersch-Wenzel and Wilhelm Treue (eds.), *Tiergarten, Teil 1: Vom Brandenburger Tor zum Zoo* (Geschichtslandschaft Berlin, Orte und Ereignisse, ii; Berlin, 1989), 323–41, here 337; and a copy of a report by Lutz Heck to the members of the supervisory board, AZGB O 0/1/54.

114 Copy of a report by Lutz Heck to the members of the supervisory board, AZGB O 0/1/54.

115 Cf. for example Gabriele Hammermann, *Zwangsarbeit für den "Verbündeten": Die Arbeits- und Lebensbedingungen der italienischen Militärinternierten in Deutschland 1943–1945* (Tübingen, 2002); Gabriele Hammermann, "Die italienischen Militärinternierten zwischen 1943 und 1945" in *Zwischen Allen Stühlen: Die Geschichte Der Italienischen Militärinternierten 1943–1945*, ed. Nazi Forced Labour Documentation Centre of the Topography of Terror Foundation (Berlin, 2017), 28–41; Daniela Geppert and Arne Pannen, "Italienische Militärinternierte im Kriegsgefangenenlager der Reichshauptstadt" in ibid., 52–61.

116 Schürer and Schmerenbeck, *100 Jahre Zoo Wuppertal*, 66; Mustafa Haikal and Jörg Junhold, *Auf der Spur des Löwen: 125 Jahre Zoo Leipzig* (Leipzig, 2003), 156; Nastasja Klothmann, *Gefühlswelten im Zoo* (Bielefeld, 2015), 185; Evelyn Gottschlich, "Wiedereröffnung und Zerstörung", Helmut Zedelmaier and Michael Kamp (eds.), *Nilpferde an der Isar: Die Geschichte des Tierparks Hellabrunn in München* (Munich, 2000), 112–38, here 137.

117 Letter from Zoologischer Garten Berlin to the Wehrmacht welfare officer in Berlin C, 24 Mar. 1945, AZGB O 0/1/43.

118 Zoologischer Garten Berlin annual report for 1943, AZGB S 14/2.
119 Gertrud Haupt, "Schimpansenstation im splittersicheren Kellergeschoß", *Das 12-Uhr-Blatt*, 18 Jan. 1944.
120 Annual report for 1943, AZGB S 14/2.
121 Zoologischer Garten Berlin old animal register.
122 Copy of a report by Lutz Heck to the members of the supervisory board, AZGB O 0/1/54.
123 Copy of a letter from L. Heck to O. Antonius, 19 July 1943, AZGB 5/23.
124 Werner Girbig, *Im Anflug auf die Reichshauptstadt: Die Dokumentation der Bombenangriffe auf Berlin* (Stuttgart, 2001), 109.
125 Zoologischer Garten Berlin old animal register and copy of a report by Lutz Heck to the members of the supervisory board, AZGB O 0/1/54.
126 Copy of a report by Lutz Heck to the members of the supervisory board, AZGB O 0/1/54.
127 Recollections of Elisabeth Johst, AZGB S 15/27.
128 Lutz Heck, Der Luftangriff vom 22. November 1943 auf den Berliner Zoologischen Garten, AZGB N 4/5.
129 Copy of a report by Lutz Heck to the members of the supervisory board, undated, AZGB O 0/1/54; Heck, L., Der Luftangriff vom 22. November 1943 auf den Berliner Zoologischen Garten, AZGB N 4/5; bomb damage register, "Straussenhaus" index card AZGB O 0/1/54.
130 Maria von Maltzan, *Schlage die Trommel und fürchte dich nicht* (Berlin, 1986), 158.
131 O. Heinroth, personal report on the bomb damage at Aquarium and Zoo in the nights of 22 and 23 Nov. 1943, the estate of Oskar Heinroth, manuscripts department of the Berlin State Library.
132 Reinhard Einsel, "Erinnerungen an den Berliner Zoo", *Bongo* 14 (1988), 83–6.
133 T. C. Lamers, "Die Nacht, als das Aquarium in Schutt und Asche fiel", *Bongo* 33 (2003), 25–9.
134 Entry for 14 Dec. 1943 in the diary of Joseph Goebbels, *Die Tagebücher von Joseph Goebbels, Teil II*, ed. Elke Fröhlich for the Institute for Contemporary History Munich (Munich, 1994), 476.
135 Zoologischer Garten Berlin annual report for 1943, AZGB S 14/2.
136 Copy of a report by Lutz Heck to the members of the supervisory board, AZGB O 0/1/54, also quoted in a transcript of the defence plea submitted by lawyer Fritz Detert, 26 Mar. 1957, AZGB O 0/1/47.
137 Zoologischer Garten Berlin old animal register.
138 Transcript of the supervisory board meeting on 25 Oct. 1944, AZGB O 0/2/6.
139 Memo by L. Heck regarding the reopening on 25 July 1944, AZGB O 0/1/50; recollections of Elisabeth Johst, AZGB S 15/27; carbon copy of a letter from L. Heck to Tiergrotten und Nordsee-Aquarium Wesermünde, 1 July 1944, AZGB O 0/1/40.
140 Lutz Heck, "Hundert-Jahr-Feier am 1. August 1944" and "Kurzer Bericht", AZGB N 5/13.
141 Burghard Ciesla and Helmut Suter, *Jagd und Macht: Die Geschichte des Jagdreviers Schorfheide* (Berlin, 2011), 189.
142 Supervisory board meeting minutes, 25 Oct. 1944, AZGB O 0/2/6; transcript of a letter from L. Heck to Otto Antonius (Schönbrunn Zoo), 13 Oct. 1944, AZGB O 0/1/40.
143 Katharina Heinroth, *Mit Faltern begann's: Mein Leben mit Tieren in Breslau, München und Berlin* (Munich, 1979), 139.
144 Staff index card for Lutz Heck, AZGB, N 4/4.
145 Lutz Heck, "Gründe meines Verlassens der Stadt Berlin am 7.5.1945", AZGB N 5/13; Susanne Linden, "The Zoological Gardens in the Tiergarten", University of Ohio Library, Cornelius Ryan Collection of World War II Papers, Box 59, Folder 31.
146 Linden, "The Zoological Gardens".
147 Recollections of Elisabeth Johst, AZGB S 15/27k.
148 For this information I am indebted to Hans-Albert Hoffmann.
149 Interview with Elisabeth Johst, University of Ohio Library, Cornelius Ryan Collection of World War II Papers, Box 59, Folder 34; recollections of Elisabeth Johst, AZGB S 15/27.
150 K. Heinroth, "Kriegszerstörungen und Aufbau von 1945 bis 1956 im Berliner Zoologischen Garten", typewritten manuscript, AZGB N 4/2.
151 K. Heinroth to W. Keller, 18 Sept. 1945, AZGB O 0/1/87; K. Heinroth to G. Freytag, 3 Jan. 1946, AZGB O 0/1/86; K. Heinroth, "Professor Heck und der Zoo", "Democratic Forum" segment, *Tagesspiegel*, 23 May 1954.
152 Carbon copy of a letter from L. Heck to the army uniform stores in Berlin, 9 Mar. 1945, AZGB O 0/1/41; L. Heck to K. Heinroth, 19 Nov. 1945, AZGB N 4/4.
153 Interview with Lutz Heck, University of Ohio Library, Cornelius Ryan Collection of World War II Papers, Box 59, Folder 32.
154 L. Heck: "Gründe meines Verlassens der Stadt Berlin am 7.5.1945", 1951, AZGB N 5/13.
155 On 11 Nov. 1945, zoologist and bison expert Erna Mohr wrote to K. Heinroth that Heck was wanted for the destruction of Askania-Nova, AZGB O 0/1/87. Kai Artinger agreed with this interpretation in Artinger, *Lutz Heck*, footnote 65.

156 Recollections of Elisabeth Johst, AZGB S 15/27.
157 Heinroth, *Mit Faltern*, 139 ff.; Elisabeth Johst also confirmed this, AZGB S 15/27.

Emerging from the ruins, 1945–1949
(pp. 133–145)

1 Excerpt from Konstantin Simonow's report in Hans-Norbert Burkert, Klaus Matussek and Doris Obschernitzki, *Zerstört, besiegt, befreit: Der Kampf um Berlin bis zur Kapitulation 1945* (Stätten der Geschichte Berlins, vii; Berlin, 1985), 198.
2 Bomb damage register, "Nilpferdhaus" index card, AZGB O 0/1/54.
3 "Alo", "Pongo – das Urviech Europas", *Nacht-Depesche*, 27 Nov. 1952, 3.
4 Unknown author, activity report of the Zoologischer Garten Berlin from 3 to 23 May 1945 and lists of those buried, AZGB O 0/1/75.
5 Recollections of Willi Kettner, AZGB S 15/26.
6 Recollections of Elisabeth Johst, AZGB S 15/27.
7 Katharina Heinroth, *Mit Faltern begann's: Mein Leben mit Tieren in Breslau, München und Berlin* (Munich, 1979), 136–45.
8 Invitation to the burial, AZGB N 11/1.
9 Animal inventory of 31 May 1945, AZGB O 0/1/54.
10 Carbon copy of a letter from K. Heinroth to H. Junker, 3 Jan. 1946, AZGB O 0/1/107, as well as carbon copies of letters from K. Heinroth to G. Karg, 20 Nov. 1945, to W. Keller, 18 Sept. 1945, and to the Paul Meyer Verlag publishing house, 22 Nov. 1946, AZGB O 0/1/87.
11 Heinz-Sigurd Rathel, "Das kleine Raubtierhaus des Berliner Zoos (1865–1973) und Erinnerungen an seine Bewohner", *Bongo* 44 (2013), 143–56, here 147.
12 Susanne Linden, "The Zoological Gardens in the Tiergarten", University of Ohio Library, Cornelius Ryan Collection of World War II Papers, Box 59, Folder 31.
13 Linden, "The Zoological Gardens".
14 Animal inventory of 31 May 1945, AZGB N 5/18.
15 Zoo Berlin to General Berzarin, 28 May 1945, AZGB S 15/17.
16 Recollections of Elisabeth Johst, AZGB S 15/27.
17 Katharina Heinroth and Werner Schroder, *Der Zoologische Garten Berlin: Erster Bericht und Wegweiser nach dem Kriege* (Berlin, 1951).
18 L. Heck to the shareholders' association, 18 Nov. 1945, AZGB N 4/4; K. Heinroth to W. Keller, 18 Sept. 1945, AZGB O 0/1/87.
19 Recollections of Elisabeth Johst, AZGB S 15/27.
20 Memo on discussions with the deputy mayor of the Tiergarten district on 25 May and 7 June 1945, AZGB O 0/1/50.
21 Transcript of public meeting of the 3rd civil senate of the superior court of Berlin, 29 Jan. 1949, on the legal dispute between the Zoo and the circus with witness statements from Übrick, Knopka, Schirrmacher and Bachmann, as well as a sworn statement by Fritz Bachmann, 6 Sept. 1948, and a sworn statement by Ernst Übrick, 14 July 1948 in AZGB O 0/1/124.
22 Heinroth, *Mit Faltern*, 147 ff.
23 Handwritten journal of the Zoological Garden, 1945–1948.
24 Animal inventory of Zoologischer Garten Berlin on 18 Sept. 1945; Allied Kommandatura Berlin, Colonel Goncharov to the Governing Mayor of Berlin, 10 Nov. 1945, LAB C Rep. 101, No. 45.
25 Report (probably by Werner Schröder himself) during Schröder's time as a management board member, undated, AZGB S 15/12.
26 Transcript at the behest of Mr Weidehof, by order of the Governing Mayor, 28 July 1945, AZGB O 0/1/75.
27 Report of a Dr Setegast to the Magistrate's museum department, 24 July 1945, AZGB O 0/1/75.
28 Activity report, K. Heinroth and W. Schröder, 12 Sept. 1945, AZGB O 0/1/75; appointment of the new supervisory board and correspondence on the inaugural meeting, AZGB N 11/1.
29 Recollections of Willi Kettner, AZGB S 15/26.
30 LAB C-Rep 375/01/08, No. 9011, sheets 1–3; for information about Steinmetz, who became a member of the NSDAP on 1 Mar. 1933, see also: Federal Archives, R 9361/II, Steinmetz, Hermann, 23 Aug. 1898.
31 K. Heinroth to the Joachimsthal forestry administration, 9 Oct. 1945, AZGB O 0/1/86.
32 "An alle Berliner Tierfreunde!", *Berliner Zeitung*, 9 June 1945.
33 Handwritten journal of the Zoological Garden, 1945–1948.
34 Animal inventory of 31 Dec. 1945, AZGB O 0/1/124.
35 K. Heinroth to G. Karg, AZGB O 0/1/87.
36 Death certificate, AZGB O 0/1/44; recollections of Elisabeth Johst, AZGB S 15/27.
37 The Bittner company to the Zoo, 28 Aug. 1946, AZGB O0/1/44.
38 K. Heinroth to K. Berthold, 30 Nov. 1945, AZGB O 0/1/44.
39 Carbon copy of a letter from K. Heinroth to the administrative director of the Berlin city forests, 20 Feb. 1946, AZGB O 0/1/44; in this letter Heinroth asked for the allocation of land for the cultivation of animal fodder.
40 K. Heinroth, memo, 19 Mar. 1946 (about a conversation with the British Colonel Mander), AZGB O 0/1/75.

41 Carbon copy of a letter from K. Heinroth to Gustav Jahn, 5 Sept. 1945, and to the state forestry office in Potsdam, 20 Feb. 1946, AZGB O 0/1/87.

42 Handwritten journal of the Zoological Garden, 1945–1948; K. Heinroth to police station 31, AZGB O 0/1/143.

43 K. Heinroth to police station 31, 12 Apr. 1947, AZGB O 0/1/143.

44 Carbon copy of a letter from K. Heinroth to an unknown recipient, 19 Mar. 1946, AZGB O 0/1/86.

45 Report of a burglary, 26 June 1947; press release about a theft at the Zoo, 25 Feb. 1947, AZGB S 15/17; K. Heinroth to police station 31, 1 Nov. 1947, AZGB O 0/1/143.

46 K. Heinroth to police station 31, 31 May 1947, AZGB O 0/1/143.

47 K. Heinroth to police command, 1 Apr. 1947, AZGB O 0/1/143.

48 K. Heinroth to Tiergarten police commissioner, 10 Mar. 1947, AZGB O 0/1/143; K. Heinroth to H.-U. Mrozek, 8 Mar. 1947, AZGB O 0/1/87.

49 K. Heinroth, memo, 5 June 1946, AZGB O 0/1/124.

50 Carbon copy of a letter from Katharina Heinroth to the administrative director of the Berlin city forests, 20 Feb. 1946, AZGB O 0/1/44.

51 State Party School of the SED to the Zoo, 21 June 1946, and a copy of the answer from 25 June 1946, AZGB O 0/1/216.

52 K. Heinroth's correspondence with E. Mohr and W. Keller in 1945, AZGB O 0/1/87.

53 Carbon copy of a letter from Katharina Heinroth to Heinz Heck, 6 Aug. 1947, AZGB O 0/1/58.

54 I am indebted for this information to the former Wuppertal zoo director Ulrich Schürer.

55 Heinroth, *Mit Faltern*, 267.

56 Animal inventory 31 Dec. 1947, AZGB O 0/1/124.

57 Information on the flak towers in Berlin can be found in Michael Foedrowitz, *Die Flaktürme: Berlin – Hamburg – Wien* (Berlin, 2017).

58 Statement of costs for damage caused by the first blasting operation on 30 Aug. 1947; statement of costs for damage caused by the second blasting operation on 30 July 1948, AZGB O 0/1/75.

59 Announcement, 3 Aug. 1948, AZGB O 0/1/55.

60 Notification from the state health authority, veterinary department, 6 July 1948, AZGB S 15/17.

61 Martin Schaaff, "Paula Busch und der Berliner Zoo", *Bongo* 14 (1988), 87–92.

62 K. Heinroth to L. Beyer on 16 Dec. 1947, AZGB O 0/1/44.

Normalisation during the Cold War, 1949–1961 (pp. 147–175)

1 Werner Schröder, *Zum Abschied ein Krokodil: Heiter-Besinnliches über Mensch und Tier* (Bad Münstereifel, 1999), 129.

2 "Bulette schon populär", *Der Tag*, 1 Sept. 1953; "Knautschkes Tochter heißt Bulette", *Telegraf*, 1 Sept. 1953.

3 Katharina Heinroth and Werner Schröder, *Der Zoologische Garten Berlin: Erster Bericht und Wegweiser nach dem Kriege* (Berlin, 1951). Other examples include: Katharina Heinroth, "Kriegszerstörungen und Aufbau von 1945 bis 1956 im Berliner Zoologischen Garten", Sitzungsberichte der Gesellschaft der Naturfreunde zu Berlin N. F. (1961), 1, 48–58; Katharina Heinroth, *Mit Faltern begann's: Mein Leben mit Tieren in Breslau, München und Berlin* (Munich, 1979), 153; Bernhard Blaszkiewitz, *Knautschke, Knut & Co.: Die Lieblingstiere der Berliner aus Tierpark und Zoo* (Berlin, 2009), 12; and a press release published by Zoologischer Garten Berlin, 40/83, 6 May 1983.

4 Heinroth and Schröder, *Der Zoologische Garten Berlin*.

5 Postcard written by Hello Kellner, 16 Mar. 1947, AZGB O 0/1/87.

6 Carl Mampe AG to K. Heinroth, 6 Dec. 1949; correspondence between K. Heinroth and Paul Miehe in Sept. 1949, AZGB O 0/1/60.

7 Heinroth, *Mit Faltern*, 195ff.

8 J. Brachetka to K. Heinroth, 16 Feb. 1953, AZGB O 0/1/162.

9 "Rieke ist die erste von 1100", *Nacht-Depesche*, 21 Aug. 1953.

10 Recollections of Elisabeth Johst, AZGB S 15/27.

11 Reviews of the Oktoberfests 1954/55 and 1956, AZGB O 3/2/5; Zoologischer Garten Berlin annual reports for 1954 and 1958 (Berlin, 1955 and 1959).

12 "Schotten musizieren und tanzen", *Die Welt*, 28 June 1949.

13 Heinz-Georg Klös, *Wegweiser durch den Zoologischen Garten Berlin* (Berlin, 1959), 57 plus map; "egl.", "Der zweite Bahnhof Zoo", *Der Tagesspiegel*, 8 July 1955.

14 For the deliberations in the East Berlin Magistrate, see Siegfried Prokop, "Frühe Pläne für einen 'Volks-Tierpark' und die Gründung des Tierparks in Friedrichsfelde 1953–1955", in Kathrin Böhme, Ekkehard Höxtermann and Wolfgang Viebahn (eds.), *Heinrich Dathe: Zoologe und Tiergärtner aus Leidenschaft* (Biologiehistorische Symposien; Rangsdorf, 2015), 114–30.

15 Heinrich Dathe, *Lebenserinnerungen eines leidenschaftlichen Tiergärtners* (Munich, 2002), 212.

16 Jürgen Hofmann, "Heinrich Dathe zwischen Politik, Volksbildung und Wissenschaft", in Böhme, Höxtermann and Viebahn, *Heinrich Dathe*, 93–100, here 95; Dathe, *Lebenserinnerungen*, 86ff.

17 Guards Regiment to H. Dathe, 31 Oct. 1957, and H. Dathe to the Guards Regiment, 6 Nov. 1957, BStU, MfS, WR, Fo 0003, images 0003 and 0004.

18 "Berliner Tierparadies wird noch schöner", *Neues Deutschland*, 26 Jan. 1964.

19 Zoologischer Garten Rostock gGmbH, *Vom Hirschgarten zum Erlebniszoo: 100 Jahre Tiergarten Rostock* (Rostock, 1999), 23.

20 "ex", "Kein Wasserbecken für die Eisbären", *Der Kurier*, 4 Aug. 1955.

21 *Der Tag*, 30 June 1955.

22 Heinrich Dathe, "Willkommen im Berliner Tierpark", *Neues Deutschland*, 30 June 1955.

23 "'VEB-Kälte kaufte Eisbären", *Der Tagesspiegel*, 1 July 1955.

24 H. G. S., "Ostberlin eröffnet heute seinen eigenen Tierpark", *Die Welt*, 2 July 1955.

25 Copy of a letter from H. Dathe to H. Demmer, 2 Aug. 1958, AZGB N 5/14.

26 Copy of a letter from H. Dathe to Mayor F. Ebert, 5 Aug. 1958, AZGB O 0/1/32.

27 Zoologischer Garten Berlin annual report for 1956, AZGB O 0/3/10, and a list of visitor numbers, AZGB O 0/1/62 and N 5/18.

28 Proposals for the building department's works programme for 1949, AZGB O 0/1/75.

29 Manuscript of W. Schröder's speech on the occasion of the reopening of the Crocodile Hall, 30 Nov. 1956, and a news item, AZGB S 15/12; Werner Schröder, *Zum Abschied ein Krokodil: Heiter-Besinnliches über Mensch und Tier* (Bad Münstereifel, 1999), 65.

30 For this information I am indebted to the former director of the Wuppertal zoo, Dr Ulrich Schürer.

31 Carbon copy of a letter from Katharina Heinroth to Erwin Leser, 28 Oct. 1948, AZGB N 4/4.

32 Copy of a letter from J. Gebbing to the chair of the Association of German Zoo Directors, 29 Apr. 1954, AZGB N 5/13.

33 B. Grzimek to Katharina Heinroth, 3 Dec. 1952 and 4 Mar. 1953, AZGB N 4/6.

34 There are numerous interviews and articles in regional newspapers from this time. He probably hoped to gain a greater scientific reputation with "Über den Auerochsen und seine Rückzüchtung", *Jahrbuch des Nassauischen Vereins für Naturkunde* 90 (1952), 107–24.

35 "Wird Professor Lutz Heck in den Zoo zurückgerufen?", *Der Kurier*, 16 Mar. 1954; H. D., "Zoo mit Weltruf ", reader's letter in the *Berliner Morgenpost*, Kreuzberg edition, 31 Mar. 1954; "W", "Der Zoo soll wieder repräsentativer Tiergarten werden", *Der Kurier*, 15 Apr. 1954; "Berliner Zoo: Urmacher unerwünscht", *Der Spiegel*, 23 June 1954, 12 and 14.

36 Hans Nachtsheim, "Professor Heck und der Zoo", *Der Tagesspiegel*, 4 April 1954; "Dunkle Kaempfe um den Zoo", *Nacht-Depesche*, 22 June 1954, 3; "Yth", "Wühlarbeit gegen den Zoo", *Telegraf*, 26 June 1954.

37 K. Heinroth, "Professor Heck und der Zoo", *Der Tagesspiegel*, 23 May 1954, 1.

38 Manuscript of a report by K. Heinroth for the AGM, 30 June 1954, AZGB O 0/3/10.

39 "Der Hubschrauberlandeplatz am Zoologischen Garten Berlin", *Bauwelt* 28 (1954), 546ff.

40 Zoologischer Garten Berlin annual report for 1953, AZGB O 0/3/10.

41 Heinroth, *Mit Faltern*, 267.

42 Ludwig Baumgarten, *Chronik Zoologischer Garten Halle, Teil 1: 1901–1945* (Halle, 2001), 142.

43 Transcript of a letter from Vera Holubek to Zoologischer Garten Berlin, 16 June 1956, and carbon copy of a letter from K. Heinroth to Theodor Frenzel dated 18 June 1956, AZGB O 0/3/5.

44 Report by K. Heinroth for the AGM in 1956, AZGB O 0/3/10.

45 "Lebhafte Zoo-Hauptversammlung", *Berliner Morgenpost*, 28 June 1956; "Zoodirektorin nahm Abschied", *B.Z. Am Abend*, 30 June 1956.

46 Report by K. Heinroth for the AGM in 1956, AZGB O 0/3/10; Heinz-Georg Klös, Hans Frädrich and Ursula Klös, *Die Arche Noah an der Spree: 150 Jahre Zoologischer Garten Berlin – Eine tiergärtnerische Kulturgeschichte von 1844–1994* (Berlin, 1994), 138.

47 H.-G. Klös to Ludwig Zukowsky, 15 July 1956, AZGB N 5/17.

48 K. Heinroth to supervisory board chair Arno Weimann, 26 Sept. 1956, AZGB N 4/7.

49 Ernst Schmidt to Zoologischer Garten Berlin, 3 June 1956, AZGB O 0/3/5.

50 Walter Riek to H.-G. Klös, 5 Feb. 1959, AZGB N 5/17.

51 Transcripts of the court ruling, AZGB N 5/17.

52 "Neue Gesichter im Aufsichtsrat des Zoo", *Berlin-Kurier*, 28 Apr. 1960, 8.

53 Heinz-Georg Klös, *Von der Menagerie zum Tierparadies: 125 Jahre Zoo Berlin* (Berlin, 1969), 138.

54 Tierpark Berlin, "Arbeitsplan für 1960", 5 Mar. 1960, LAB C Rep. 121, No. 178.

55 Tierpark press release No. 98/59, 29 Dec. 1959, LAB C Rep. 121, No. 178.

56 H. Dathe to Walter Ulbricht, 23 Aug. 1956, LAB C Rep 121, No. 46.

57 H. Dathe to the East Berlin Magistrate's department of culture, 22 Jan. 1958, LAB C Rep. 121, No. 178.
58 Memo from the department of culture to the Magistrate, 5 Nov. 1960, regarding a call by Dathe the previous day, LAB C Rep 121, No. 46.
59 H. Dathe to J. Blecha, 27 Apr. 1960, LAB C Rep. 121, No. 178.
60 Note for Heinz-Georg Klös, AZGB N 5/9.
61 Zoo to the President of Berlin's House of Representatives Willy Henneberg, 11 Mar. 1959, LAB B Rep. 014, No. 2541.
62 Note by the Senator for Public Education III b A 347, 8 June 1959, LAB B Rep. 014, No. 2541.
63 Meeting minutes, 10 Jan. 1966, AZGB O 0/1/116.
64 Copy of a memo for the Governing Mayor, 8 Aug. 1983, AZGB S 15/7.

A zoo on an "island": The 1960s (pp. 177–189)

1 "Millionen-Spritze für den Zoo", *Bild*, Berlin edition, 24 Apr. 1960.
2 Zoologischer Garten Berlin annual report for 1962 (Berlin, 1963), 5.
3 Transcript of correspondence between the Zoo and the Senator for Public Education, 28 Nov. 1961, LAB B Rep. 014, No. 2541.
4 List of visitor numbers, AZGB O 0/1/123.
5 Senate Chancellery, programme for the visit of his Excellency the United States Attorney General addressed to Robert J. McCormick, 11 May 1964, AZGB O 0/1/223.
6 *Deutschlandspiegel* 39 (1958), programme broadcast from 9 Jan. 1958.
7 Ministry for Culture's literature import department to H. Dathe, 4 Nov. 1974, AZGB O 0/1/206.
8 Telegram from H. Hediger to H. Dathe, June 1963, AZGB O 0/1/207.
9 Jan Mohnhaupt, *Der Zoo der anderen: Als die Stasi ihr Herz für Brillenbären entdeckte & Helmut Schmidt mit Pandas nachrüstete* (Munich, 2017), 142.
10 Report by H.-G. Klös at the AGM on 26 June 1957, AZGB O 0/3/10.
11 "'Garten Eden' für Bären", *Berliner Morgenpost*, 23 Nov. 1966.
12 General information can be found in the annual reports for 1967 to 1969 and in the correspondence and memos relating to the outdoor bear enclosures, AZGB O 4/1/2.
13 Programme for the 125th anniversary celebrations of Zoologischer Garten Berlin; records of costs for the anniversary and the Zoo director conference; press release "Der Zoo rüstet sich zum Fest", 26 June 1969, AZGB O 0/1/90.
14 Heinz-Georg Klös, Hans Frädrich and Ursula Klös, *Die Arche Noah an der Spree: 150 Jahre Zoologischer Garten Berlin, eine tiergärtnerische Kulturgeschichte von 1844–1994* (Berlin, 1994), 161.
15 All letters can be found in AZGB V 1/2.
16 Fritz Dietrich Altmann to an unknown recipient, Jan. 1991, AZGB N 5/16.
17 Copy of a letter from H.-G. Klös to Wolfgang Puschmann, 1 Oct. 1995, AZGB N 5/14.
18 Information about the Predator House contained in the files pertaining to the opening ceremony, AZGB S 15/21.
19 "Echte Perser im Raubtierhaus", *Tagesspiegel*, 27 Apr. 1974.
20 Martin Hinrich Lichtenstein, "Der Zoologische Garten", *Vossische Zeitung*, 13 Aug. 1844; current inventory of animals at the Zoological Garden, mid-June 1845, AZGB S 9/4, 4.
21 Fifth inventory of animals at the Zoological Garden, summer 1846, AZGB S 9/3.
22 Carbon copy of a letter from K. Heinroth to the animal protection association for Berlin and surroundings, 29 Oct. 1946, and further correspondence in AZGB O 0/1/143.
23 "Zoologischer Garten", *Königlich privilegirte Berlinische Zeitung*, 29 July 1844.
24 Current inventory of animals at the Zoological Garden, mid-June 1845, AZGB S 9/4.
25 Nigel Rothfels, "Die Revolution des Herrn Hagenbeck", in Mitchell G. Ash (ed.), *Mensch, Tier und Zoo: Der Tiergarten Schönbrunn im internationalen Vergleich vom 18. Jahrhundert bis heute* (Vienna, 2008), 203–24, here 210.
26 Katharina Heinroth and Werner Schröder, *Der Zoologische Garten Berlin: Erster Bericht und Wegweiser nach dem Kriege* (Berlin, 1951).
27 Response to the motion by Kurt Zacher to vote on the introduction of a general feeding ban, 21 June 1955, AZGB O 0/3/10.
28 Katharina Heinroth, *Mit Faltern begann's: Mein Leben mit Tieren in Breslau, München und Berlin* (Munich, 1979), 174.
29 Zoologischer Garten Berlin annual report for1959 (Berlin, 1960); "ss", "Fütterungsverbot im Zoo", *Die Welt*, 31 Mar. 1960; "Ab heute: Besuchern ist das Füttern im Zoo verboten", *Bild*, Berlin edition, 31 Mar. 1960.
30 "Elefantenkuh Jambo nicht mehr zu retten?", *Der Tag*, 31 Mar. 1960, 10.
31 All letters contained in AZGB O 0/1/111, O 0/1/112.

A green oasis in the big city: The 1970s and 80s (pp. 191–209)

1 Heinz-Georg Klös and Hans Frädrich, "Zum Gedenken an Werner Schröder", *Bongo* 11 (1986), 101–3.
2 Inventory comparison, 11 Jan. 1972, AZGB O 1/8/3.
3 Description of the renovation of the older building, 1978, AZGB O 4/1/18; Jürgen Lange, "Hinter den Kulissen", in Heinz-Georg Klös (ed.), *Festschrift 70 Jahre Aquarium* (Berlin, 1983); Bernhard Blaszkiewitz, "Geschichte und Entwicklung des Zoo-Aquariums Berlin", in idem (ed.), *Picassofisch und Kompassqualle: 100 Jahre Zoo-Aquarium Berlin* (Berlin, 2013), 14–43; Jürgen Lange, "Hundert Jahre Schauaquaristik", in ibid., 44–83; and annual reports for 1979 to 1983.
4 "Ich berlinfeindlich? Das ist dummes Zeug", *Bild*, Berlin edition, 31 May 1974
5 Copy of a letter from H.-G. Klös to Helmut Kohl, 31 Aug. 1983, AZGB S 15/7.
6 Copy of a letter from Helmut Kohl to Richard von Weizsäcker, 27 Sept. 1983, AZGB O 0/1/117.
7 Correspondence with the President of the House of Representatives of Berlin Peter Rebsch regarding Kiwis from Auckland and a memo about the visit of the President of the House of Representatives, 26 Jan. 1984, AZGB O 0/1/117.
8 Memo, 15 October 1987, AZGB O 0/1/13.
9 Speeches, including in Zoologischer Garten Berlin annual report for 1984 (Berlin, 1985).
10 Report on a meeting by IM "Claus Stein", 14 Jan. 1986, BStU Archiv No. 4658 /91 II, sheet 15.
11 IM dossier for "Wolfgang Hubert" at the Lichtenberg district office, BStU Archiv No. 2955/91 I.
12 File of full-time IM boss "Jürgen Thiel", BStU Archiv No. 4711/91 II.
13 Jan Mohnhaupt, *Der Zoo der anderen: Als die Stasi ihr Herz für Brillenbären entdeckte & Helmut Schmidt mit Pandas nachrüstete* (Munich, 2017), 73ff.
14 Correspondence with the Ministry for Culture, the SED and the Magistrate, 1973–1974, AZGB O 0/1/206.
15 BStU, MfS, Lichtenberg district office 1028, 59–61.
16 BStU, MfS, BV Berlin, AIM 4697/91 II, sheets 96 and 143ff.
17 BStU Archiv No. 0031/89, sheet 62; Archiv No. 4697/91 II, sheet 63.
18 Report on a meeting with IM "Heinz Otter", 17 Feb. 1982 and 1 Sept. 1982, BStU Archiv No. 4697/91 II, 143ff., 155ff.; note, BV Berlin, Dept. II/1, 18 Jan. 1983, BStU Archiv No. 2955/91 I, sheets 242ff.
19 Memo, 24 Nov. 1989, BStU, MfS, AGMS 4504/91, sheet 225.
20 Report on a meeting with IM "Dieter Fiebig", BStU, MfS, AGMS 4504/91, sheets 109ff.
21 GMS file "Dieter Fiebig", BStU, MfS, AGMS 4504/91; IM file "Heinz Otter", BStU Archiv No. 4697/91 II.
22 Services document of IM boss "Jürgen Thiel", Oct. 1989, BStU Archiv No. 4711/91 II; report by IM "Steinbaum", 2 Oct. 1989, BStU Archiv No. 4360/91 II, sheets 224ff.
23 Copy of the application submitted to the DKLB lottery foundation and the cost calculation, AZGB O 0/1/28.
24 Memo of 27 Dec. 1982 to the DKLB foundation; report on the business trip to Dresden on 10 Aug. 1983, correspondence with VEB Elbnaturstein, AZGB O 0/1/228.
25 Copy of a letter from H.-G. Klös to E. Diepgen, 18 Sept. 1986, AZGB O 0/1/121.
26 The "animal wish list" for the Zoo's 750th anniversary celebrations, 12 Dec. 1986; Berlin state press service list of invited guests, AZGB O 0/1/45.
27 E. Diepgen to H.-G. Klös, 31 Oct. 1986, AZGB O 0/1/121.
28 Correspondence with the ambassadors of Indonesia and Malaysia, with the Chancellery and more, AZGB O 0/1/45.
29 Zoologischer Garten Berlin annual report for 1987 (Berlin, 1988).
30 Appendix to report on a meeting by GMS "Richard", 7 Mar. 1986, BStU Archiv No. 0031/89, sheet 62.
31 Report on meetings by IM "Claus Stein" on 13 and 16 Sept. 1987, BStU Archiv No. 4658/91 II, sheets 35 and 36.
32 "Tierfang", *Deutsch-Ostafrikanische Zeitung*, 3 Dec. 1904.
33 This is particularly vividly portrayed in the annual report for 1895 (Berlin, 1896), since it describes "German East Africa" and "German South West Africa" as equally valuable sources of gifts.
34 List of war damages, 15 Mar. 1945, AZGB N 2/5.
35 Heinz-Georg Klös, "Vom Umfolozi in den Berliner Zoo", annual report for 1963 (Berlin, 1964).

One city, two zoos: Zoo and Tierpark Berlin in the 1990s (pp. 211–229)

1 Zoologischer Garten Berlin AG press releases 89/89 and 90/89, 15 Nov. 1989, 21 Nov. 1989 and 13 Dec. 1989; Zoologischer Garten Berlin annual report for 1989 (Berlin, 1990).
2 Zoologischer Garten Berlin annual report for 1989 (Berlin, 1990).
3 Draft annual report for 1989, provided by Prof. H. Dathe, AZGB O 0/1/113; printed copy of the Tierpark Berlin annual report for 1989 (Berlin, 1990); Tierpark Berlin annual report for 1988 (Berlin, 1990); Zoolo-

gischer Garten Berlin annual report for 1988 (Berlin, 1989).

4 Heinrich Dathe, *Wegweiser durch den Park* 27 (Berlin, 1989).

5 Memo by H. Frädrich on his conversation with state secretary Prof. Schröder at Schöneberg city hall on 26 Oct. 1990 (strictly confidential), AZGB N 5/15.

6 IM dossier Berlin regional office XX/396/84, "Claus Stein", BStU Archiv No. 4658/91 I.

7 Senate Department for Finance to the executive of Zoologischer Garten Berlin AG, 18 Dec. 1990, AZGB N 5/15.

8 B. Blaszkiewitz, "Überlegungen zur Beziehung Zoologischer Garten Berlin/ Tierpark Berlin", 7 Oct. 1990; H. Frädrich, "Gedanken über das künftige Verhältnis zwischen dem Zoologischen Garten und dem Tierpark Berlin", 26 Oct. 1990, AZGB O 0/1/131.

9 Supervisory board meeting minutes, 23 Nov. 1990, AZGB N 5/15.

10 Speech by Prof. Klös on the occasion of the 80th birthday of Prof. Dathe, 7 Nov. 1990, AZGB N 5/15.

11 Cf. reports on meetings by IMs "Heinz Otter" (BStU Archiv No. 4697/91 II) and "Wolfgang Hubert" (BStU Archiv No. 2955/91 I).

12 Copy of template letter by Finance Senator N. Meisner; Finance Press Office: Tierpark press release "Aus ist eine Ente!", 14 Dec. 1990; press release provided by Meisner: "235 Mitarbeiter des Tierparks erhalten dauerhafte Arbeitsplätze", 22 Dec. 1990, AZGB N 5/15.

13 Senate Department for Finance press release "Existenz des Tierparks gesichert", 26 Mar. 1991.

14 Statement on snake-keeping by Jürgen Lange, 1990, AZGB N 5/10.

15 J. Lange to H.-G. Klös, 11 Aug. 1993, AZGB N 5/12

16 J. Lange to H.-G. Klös, 26 July 1993, AZGB N 5/12

17 Minutes of the joint meeting, 15 Sept. 1993; Public Services, Transport and Traffic Union (ötv) to the Tierpark-Friedrichsfelde supervisory board, Nov. 1993; Zoologischer Garten Berlin works council to H.-G. Klös, 15 Oct. 1993, AZGB O 0/2/15.

18 Helmut Höge, "Rettet die Ossifanten!", *Die Zeit*, 29 Oct. 1993; "Widerstand gegen Anfang vom Ende des Tierparks", *Neues Deutschland*, 13 Oct. 1993.

19 Supervisory board meeting minutes, 4 Nov. 1993, AZGB O 0/2/15.

20 Carbon copy of a letter from J. Lange to Christa Krey, 2 Aug. 1990, AZGB O 1/2/32 (which also includes all further correspondence).

21 Rainer Kaiser and Jürgen Lange, "Quallenzucht – ein weltweites Zuchtprogramm", *Bongo* 33 (2003), 38–44.

22 Jürgen Lange, Tai Motofumi and Rainer Kaiser, "Husbandry of jellyfish, from the beginning until today", *Der Zoologische Garten* 85 (2016), 52–63; Rainer Kaiser, "Aquarium für Quallen, ein eingetragenes Patent des Zoo-Aquarium Berlin", *Bongo* 41 (2011), 67–74.

Modern challenges: The Zoo in the new millennium (pp. 231–243)

1 Zoologischer Garten Berlin AG press release, 5 Oct. 2000, application and correspondence, AZGB O 0/1/214.

2 I am indebted to Zoo veterinarians Andreas Ochs and André Schüle and curators Heiner Klös and Ragnar Kühne for the long, friendly conversations that provided me with detailed information about Knut's life.

3 Zoologischer Garten Berlin AG press releases 04/07, 22 Jan. 2007, and 15/07, 28 Feb. 2007.

4 Zoology department of Zoologischer Garten Berlin AG, daily reports book Mar. 2007.

5 Press release by Rudolf Reinhard, Zoologischer Garten Berlin AG, 44/07, 4 July 2007.

6 Zoologischer Garten Berlin AG and Universal Music press release 25/07, 17 Apr. 2007.

7 Sigmar Gabriel, "Kann Knut den Klimawandel bremsen?", *B.Z.* special supplement: "Das große Knut-Album" (Berlin, 2007), 49.

8 *Knut und seine Freunde*, dir. Michael G. Johnson, prod. DOKfilm Fernsehproduktion GmbH, Norddeutscher Rundfunk, Rundfunk Berlin Brandenburg, 2008.

9 Zoologischer Garten Berlin AG and Haribo GmbH press release, 28 Sept. 2007.

10 Press release published by Zoologischer Garten Berlin AG, 21/2011, 20 Mar. 2011; Leibniz Institut für Wildtierforschung, "Rätsel um Eisbär Knuts Erkrankung gelöst", 27 Aug. 2015, http://www.izw-berlin.de/presse mitteilung/raetsel-um-eisbaer-knutserkrankung-geloest.html.

A Zoo fit for the future (pp. 245–259)

1 Steffi Kammerer, "Im Gehege des Vergessens", *Süddeutsche Zeitung*, 12 Oct. 2000.

2 Monika Schmidt, "Die 'Arisierung' des Berliner Zoologischen Gartens", *Jahrbuch für Antisemitismusforschung* 12 (Berlin, 2003), 210–29.

3 Press release published by Zoologischer Garten Berlin AG, 43/11, 27 June 2011.

4 Kai Artinger, "Lutz Heck: Der 'Vater der Rominter Ure'. Einige Bemerkungen zum wissenschaftlichen Leiter des Berliner Zoos im Nationalsozialismus", *Der Bär von Berlin: Jahrbuch des Vereins für die Geschichte Berlins* 23 (1994).

5 Speeches by Dietrich von Grunelius and Heinz-Georg Klös on the occasion of the bust's unveiling, AZGB S 15/18.
6 Press release published by Zoologischer Garten Berlin AG "Blick in die Vergangenheit des Zoos", 1 Dec. 2016.
7 Copy of a memo for the Governing Mayor of Berlin, 8 Aug. 1983, AZGB S 15/7.
8 Heinz-Georg Klös, "Jahresbericht für 1980", *Bongo* 5 (1981), 97–168.
9 Ferdinand Damaschun, "Tien Tien", in idem (ed.), *Panda: Das Buch zur Sonderausstellung im Museum für Naturkunde* (Berlin, 2015), 75.
10 Copy of a letter from Horst Teltschick to Heinz-Georg Klös, 31 Oct. 1984, copy of a letter from E. Diepgen to Chinese premier Zhao Ziyang, 1 Nov. 1985, copy of a letter from Ambassador Per Fischer to Consul General Bruno Schubert, 12 Feb. 1987, and various memos from 1984 to 1986 in AZGB O 0/1/121.
11 Ferdinand Damaschun (ed.), *Panda: Das Buch zur Sonderausstellung im Museum für Naturkunde* (Berlin, 2015), with particular reference to the chapter on politics and the "biographies" of the Berlin pandas.
12 Andreas Ochs, "Bemerkungen zur Krankengeschichte des Großen Pandas (Ailuropeda melanoleuca) 'Bao Bao' im Zoo Berlin", *Bongo* 44 (2013), 23–34.
13 Press release published by Zoologischer Garten AG, 31 Oct. 2016.
14 Press release published by Zoologischer Garten AG, 21 Feb. 2017.
15 Information at: www.silentforest.eu.

Archives

Archives of Zoologischer Garten Berlin AG (AZGB)
Archives of RA Jürgen Naatz (lawyer)
Bundesbeauftragter für die Unterlagen des Staatssicherheitsdienstes der ehemaligen Deutschen Demokratischen Republik (BStU) – Commissioner for the Records of the State Security Service of the former German Democratic Republic
Bundesarchiv Berlin-Lichterfelde (BArch) – German Federal Archives
BVG Archives
Dokumentationszentrum NS-Zwangsarbeit – Nazi Forced Labour Documentation Centre
Geheimes Staatsarchiv Preußischer Kulturbesitz (GStA PK) – Prussian Secret State Archives
Kreismuseum Wewelsburg – district museum in Westphalia
Landesarchiv Berlin (LAB) – Berlin State Archives
Historische Bild- und Schriftgutsammlungen am Museum für Naturkunde – Historical image and document collections of the Berlin Museum für Naturkunde
State Archives, Freiburg
Stiftung Preußische Schlösser und Gärten Berlin-Brandenburg (SPSG) – The Prussian Palaces and Gardens Foundation Berlin-Brandenburg
Tierpark Berlin-Friedrichsfelde GmbH (photo archive)
Zoologischer Garten Berlin AG (photo collection)

Picture credits

Allgemeine Bauzeitung, XXXVI (1902), p. 33

Archives of Zoologischer Garten Berlin (AZGB): p. 16 right, 25 (Peter Joseph Lenné), 26/27, 28, 29, 30, 32 (Simon Schropp & Comp.), 34, 37 top and bottom, 38 top and bottom, 39 left, 39 right (Heinrich von der Becke), 42 left, 42 right (Friedrich Seidenstücker), 43 bottom, 44, 45, 47, 48 top and bottom, 50 top and bottom, 51 top and bottom, 52, 53, 55 top and bottom (left–right), 57, 58 top and bottom, 59, 60 bottom, 62, 63 left and right, 64, 65, 69 top and bottom, 70 top and bottom, 71 top and bottom, 72, 73 bottom, 75, 76 top and bottom, 77 left and right, 78 top, bottom and right, 80 bottom, 83, 84, 85 left (Richard Horlemann), 85 top right (Friedrich Seidenstücker), 85 bottom right, 86 top and bottom, 87, 89 (Lutz Heck), 90, 91 top left, bottom and right, 95 right and left, top and bottom, 96 bottom, 99, 101, 102 top, middle and bottom, 103, 104 top and bottom, 105, 106, 107, 108, 112 left and right, 113, 114, 115, 119, 121, 123, 124, 125, 126, 127, 128 top, 128 bottom (Friedrich Seidenstücker), 129, 130, 136, 137 (Friedrich Seidenstücker), 138, 139, 140, 141, 143 (Heinrich von der Becke), 145, 148 bottom, 149 (Kurt Hoppe), 150 left and right, 151, 152, 153 top (Irmgard Zielinski), 153 bottom, 154 top and bottom, 155 (A. Lassberg), 156 top left (Kurt Hoppe), 156 bottom left (A. Lassberg), 156 right, 157 top and bottom, 159, 160, 161 top, 161 bottom (Gerhard Budich), 163 top, 163 middle (A. Lassberg), 163 bottom (Kurt Hoppe), 164, 167, 168, 170 (Kurt Hoppe), 171 left and right, 173 right (Winfried Zellmann), 174 (Harry Porikys), 175 left and right, 175 bottom (Michael Barz), 179 top, 179 bottom (Gerhard Budich), 180, 181 top (Jurisch), 182 (Magirus), 183, 184 top and bottom, 186, 187 (Anny Grassmann), 188, 189 left, right and bottom, 193 left and right, 197 top and bottom, 198 left (Fred Kleinschmidt), 199 top, 200, 203 (Michael Barz), 205 (Fred Kleinschmidt), 207, 208 (Lutz Heck), 209 top and bottom, 213 (Fred Kleinschmidt), 214, 219, 221, 222, 223, 225 (Martin Brock), 227 (Martin Brock), 228 top and bottom (Martin Brock, 229 (Martin Brock), 230, 232, 233 top and bottom, 234, 235, 236 (Karl Bröseke), 237, 238 (Karl Bröseke), 239 (Peter Griesbach), 240 (Karl Bröseke), 241, 242 (Karl Bröseke), 243 (Karl Bröseke), 248, 249 top and bottom (Fred Kleinschmidt)

Archives of RA Jürgen Naatz (lawyer): 98

British Army Public Information, Photo Section: p. 181 bottom (Henrik Pastor)

Daheim, No. 29, 1879: p. 60 top; No. 6, 1882: p. 61

Der Bazar, 15 April1882: p. 56

Die Gartenlaube, No. 16, 1873: p. 12/13

Nazi Forced Labour Documentation Centre: p. 121

Carsten Haas: p. 16 left

Ruben Gralki: p. 8, 10 left

Illustrirte Zeitung, 27 September 1873, illustration by H. Leutemann: p. 46

Günther Klebes, *Die Straßenbahnen Berlins in alten Ansichten*, Zaltbommel 1981, p. 54

Kreismuseum Wewelsburg: p. 96 top (Inv No. 16572)

Landesarchiv Baden-Württemberg, Freiburg: p. 117 (F 176/13 No. 1424-14a-3)

Landesarchiv Berlin: p. 80 top

Leineweber GmbH & Co./Brax: p. 173 left

Leipziger Illustrirte Zeitung, 4 May 1872: p. 43 top; 26 October 1899: p. 49; 19 March1903: p. 73 top; 8 May 1869: p. 74 bottom

Metropol-Verlag: p. 247

Museum für Naturkunde – Leibniz Institute for Evolution and Biodiversity Science, historical image and document collections: p. 19 bottom, 21, 36 left

Private ownership: p. 36 right

Patrik Rösler: p. 254

Joel Sartore / National Geographic Photo Ark: p. 14

Stiftung Preußische Schlösser und Gärten Berlin-Brandenburg: p. 18 (SPSG, PK 3553, illustration by Gerhard Koerber, photography by Daniel Lindner)

Schönbrunn Zoo, historic collections: p. 17

WWF/Emil Khalilov: p. 258

Zoo Berlin: p. 10 right, 16 right, 22, 24 top and bottom, 40 (Michael George), 66, 68 top and bottom (Martin Brock), 92, 94, 132, 134 top and bottom (Peter Griesbach), 146 (Martin Brock), 148 top, 176, 178 (Heiko Lehmann), 190 (Marc Scharping), 192 (Marc Scharping), 195 (Paco de la Luz), 199 bottom, 210 (Heiko Lehmann), 212 (Karl Bröseke), 226 (Peter Griesbach), 230, 244 (Heiko Lehmann), 246, 250, 251 (Frederic Schweizer), 253, 255, 257 top (Frank Roesner / dan pearlman), 257 bottom (Frederic Schweizer), 259 (Henrik Gergen), 262 top, left and right

Zoological Society of London (ed.), The Zoological Keepsake, or, Zoology, and the Garden and Museum of the Zoological Society (London, 1830): p. 19 top

The Author

Dr Clemens Maier-Wolthausen
Historian Clemens Maier-Wolthausen was born in 1973. He has worked at the European University Institute in Florence, the Graduate School at Lund University, the Institute for the History of the German Jews in Hamburg and the Center for Research on Antisemitism (ZfA) of the Technische Universität Berlin, as well as in the German Bundestag. He has also been involved in exhibition and publication projects at institutions including Centrum Judaicum in Berlin, the German Historical Museum, the Leo Baeck Institute in New York and various German zoos. Publications include *Zuflucht im Norden: Die schwedischen Juden und die Flüchtlinge 1933–1941* (Wallstein 2019) and *Alphamännchen und hohe Tiere: Deutsch-deutsche Beziehungen in Tierpark und Zoo Berlin* (Reimer 2022).